Lena Kritten

Time dependent tomography by balloon-borne spectroscopy

Lena Kritten

Time dependent tomography by balloon-borne spectroscopy

Profiling of UV/vis absorbing radicals by balloon-borne spectroscopic Limb measurements and implications for stratospheric photochemistry

Südwestdeutscher Verlag für Hochschulschriften

Impressum/Imprint (nur für Deutschland/ only for Germany)

Bibliografische Information der Deutschen Nationalbibliothek: Die Deutsche Nationalbibliothek verzeichnet diese Publikation in der Deutschen Nationalbibliografie; detaillierte bibliografische Daten sind im Internet über http://dnb.d-nb.de abrufbar.

Alle in diesem Buch genannten Marken und Produktnamen unterliegen warenzeichen-, marken- oder patentrechtlichem Schutz bzw. sind Warenzeichen oder eingetragene Warenzeichen der jeweiligen Inhaber. Die Wiedergabe von Marken, Produktnamen, Gebrauchsnamen, Handelsnamen, Warenbezeichnungen u.s.w. in diesem Werk berechtigt auch ohne besondere Kennzeichnung nicht zu der Annahme, dass solche Namen im Sinne der Warenzeichen- und Markenschutzgesetzgebung als frei zu betrachten wären und daher von jedermann benutzt werden dürften.

Verlag: Südwestdeutscher Verlag für Hochschulschriften Aktiengesellschaft & Co. KG
Dudweiler Landstr. 99, 66123 Saarbrücken, Deutschland
Telefon +49 681 37 20 271-1, Telefax +49 681 37 20 271-0
Email: info@svh-verlag.de
Zugl.: Heidelberg, Uni, Diss., 2009

Herstellung in Deutschland:
Schaltungsdienst Lange o.H.G., Berlin
Books on Demand GmbH, Norderstedt
Reha GmbH, Saarbrücken
Amazon Distribution GmbH, Leipzig
ISBN: 978-3-8381-1546-7

Imprint (only for USA, GB)

Bibliographic information published by the Deutsche Nationalbibliothek: The Deutsche Nationalbibliothek lists this publication in the Deutsche Nationalbibliografie; detailed bibliographic data are available in the Internet at http://dnb.d-nb.de.

Any brand names and product names mentioned in this book are subject to trademark, brand or patent protection and are trademarks or registered trademarks of their respective holders. The use of brand names, product names, common names, trade names, product descriptions etc. even without a particular marking in this works is in no way to be construed to mean that such names may be regarded as unrestricted in respect of trademark and brand protection legislation and could thus be used by anyone.

Publisher: Südwestdeutscher Verlag für Hochschulschriften Aktiengesellschaft & Co. KG
Dudweiler Landstr. 99, 66123 Saarbrücken, Germany
Phone +49 681 37 20 271-1, Fax +49 681 37 20 271-0
Email: info@svh-verlag.de

Printed in the U.S.A.
Printed in the U.K. by (see last page)
ISBN: 978-3-8381-1546-7

Copyright © 2010 by the author and Südwestdeutscher Verlag für Hochschulschriften Aktiengesellschaft & Co. KG and licensors
All rights reserved. Saarbrücken 2010

Zeitabhängige Profilmessung von UV/vis absorbierenden Radikalen mittels ballongestützter spektroskopischer limb Messungen und Schlußfolgerungen für die stratosphärische Photochemie

Stickstoffverbindungen spielen schon heute eine tragende Rolle beim katalytischen Abbau von stratosphärischem Ozon, und aktuelle Studien zeigen, dass ihre Bedeutung in Zukunft wachsen wird. Die hier vorgestellten ballongestützten, spektroskopischen Beobachtungen der zeitlichen und räumlichen Variation von O_3, NO_2, BrO und HONO erlauben neue Einsichten in die NO_x und NO_y Fotochemie der tropischen oberen Troposphäre sowie der unteren und mittleren Stratosphäre.

Eine neue, an die Messung von sich zeitlich ändernden Radikalen angepasste Methode der Auswertung wird vorgestellt. Sie beinhaltet spektroskopische Beobachtungen von limb gestreutem Himmelslicht unter verschiedenen Blickwinkeln in Kombination mit Strahlungstransportmodellierung und mathematischer Inversion auf einem diskreten Zeit- und Höhengitter. Die Ergebnisse der Methode werden überprüft durch einen Vergleich mit in-situ Ozon Sonden, O_3, NO_2 und BrO Messungen mittels direktem Sonnenlicht und Beobachtungen des ENVISAT/SCIAMACHY Satelliten Instruments während zeit- und ortsnaher Messungen. Diese Vergleiche zeigen die Stärke und Gültigkeit unseres Verfahrens, das meteorologische und photochemische Korrekturen der gemessenen Konzentrationen, die bisher auf Grund zeitlicher Abweichungen der zu vergleichenden Messung angewandt wurden, überflüssig macht.

Die Daten werden, exemplarisch anhand der N_2O_5 Photolyserate, weiter untersucht, um Parameter in-situ zu testen, die für die stratosphärische Ozonchemie ausschlaggebend sind. Die vorgestellte Studie deutet, verglichen mit der üblicherweise aus JPL-2006 zitierten, eine leicht größere Photolyserate an. Innerhalb der Fehlergrenzen stimmen sie aber beide überein. Schließlich wird die Beobachtung von HONO in der oberen tropischen Troposphäre unter Einbeziehung der bekannten Photochemie und der beobachteten Bildung von NO_x durch Gewitterblitze diskutiert.

Time dependent profiling of UV/vis absorbing radicals by balloon-borne spectroscopic limb measurements and implications for stratospheric photochemistry

Nitrogen bearing compounds play an important role in catalytic loss of stratospheric ozone and, as studies indicate, will become even more important in future. Here balloon-borne limb measurements of the time and altitude dependent variation of O_3, NO_2, BrO and HONO are presented, providing new insight into the NO_x and NO_y photochemistry of the tropical upper troposphere, lower and middle stratosphere. A new method is discussed aiming at the retrieval of the diurnal variation of UV/vis absorbing radicals from balloon-borne limb scattered skylight observations in a self consistent manner. The method employs the spectroscopic measurements in combination with radiative transfer modeling and a mathematical inversion on a regularized time and height grid. The retrieval is tested by comparing the results to in-situ ozone sonding, simultaneous O_3, NO_2 and BrO direct sun observations, performed on the same payload, and to measurements of the ENVISAT/SCIAMACHY satellite instrument during a collocated overpass. The comparison demonstrates the strength and validity of our approach which renders meteorological and photochemical corrections of measured radical concentrations due to temporal mismatches of corresponding observations unnecessary.

The collected data are further explored to in-situ test photochemical parameters, critical for stratospheric ozone, exemplarily for the N_2O_5 photolysis rate. The present study indicates a slightly larger N_2O_5 photolysis rate than the commonly referred JPL-2006, but agrees within the given error bars of a factor of 2. Finally, first detection of HONO in the tropical upper troposphere is reported, and discussed with respect to the known photochemistry and formation of NO_x in nearby thunderstorms.

Contents

Abstract   5

Introduction   11

1 Atmospheric dynamics and photochemistry   13
    1.1 Vertical structure of the atmosphere 13
    1.2 Atmospheric composition 15
    1.3 Dynamics in the troposphere 17
    1.4 Dynamics in the stratosphere 18
    1.5 Stratospheric - tropospheric exchange: The tropical tropopause layer 20
        1.5.1 Convection and lightning 22
    1.6 Chemistry of HONO in the troposphere 24
    1.7 Photochemistry of ozone in the stratosphere 25
    1.8 Photochemistry of nitrogen in the stratosphere 27
    1.9 Photochemistry of halogens in the stratosphere 30

2 Physics of radiation and molecular absorption   35
    2.1 Radiative transfer in the Earth's atmosphere 35
        2.1.1 Scattering 36
        2.1.2 Absorption and emission 38
        2.1.3 The equation of radiative transfer 39
    2.2 Molecular absorption in the atmosphere 40

3 Instrumental design and performance   43
    3.1 Design 43
        3.1.1 Wavelength dependency of the FOV 45

	3.2	Performance	46
		3.2.1 Instrumental noise	46
		3.2.2 Observation geometry	48
		3.2.3 Flight preparations	49
4	**Retrieval methods**		**53**
	4.1	Spectroscopic analysis	54
		4.1.1 The DOAS forward model	56
		4.1.2 Characterization of the spectral retrieval and error analysis	57
		4.1.3 DOAS analysis of the particular gases	60
	4.2	Radiative transfer modeling	64
		4.2.1 RTM McArtim	64
	4.3	Retrieval of the elevation α	65
	4.4	Retrieval of trace gas profiles	68
		4.4.1 Time weighting	69
		4.4.2 The combined kernel	70
		4.4.3 Inverse method - optimal estimation	71
		4.4.4 Other methods to constrain the retrieval	74
		4.4.5 Characterization of the profile retrieval and error analysis	77
	4.5	The retrieval of diurnal variation/chemical information	81
5	**Results and Discussion**		**83**
	5.1	Observations from aboard MIPAS-B gondola on June 13, 2005	84
		5.1.1 Flight conditions	85
		5.1.2 Measured ΔSCDs from MIPAS-B flight	86
		5.1.3 Retrieved concentration profiles from MIPAS-B flight	91
		5.1.4 Comparison of measured O_3 from MIPAS-B flight with in-situ ozone sonding	100
		5.1.5 Detection of lightning NO_x and HONO during the MIPAS-B flight	101
	5.2	Observations from aboard LPMA/DOAS gondola on June 17, 2005	104
		5.2.1 Flight conditions	105
		5.2.2 Measured ΔSCDs from LPMA/DOAS flight	106
		5.2.3 Aerosol extinction profile	110
		5.2.4 Retrieved concentration profiles from LPMA/DOAS flight	111

	5.2.5	Cross validation with direct sunlight DOAS measurements	118
5.3		Observations from aboard LPMA/IASI gondola on June 30, 2005.	121
	5.3.1	Flight conditions	121
	5.3.2	Measured ΔSCDs from LPMA/IASI flight	121
	5.3.3	Retrieved concentration profiles	125
	5.3.4	Cross validation with satellite measurements	132
	5.3.5	The diurnal variation of NO_2	134

Conclusion   **139**

Appendix   **143**

Bibliography   **154**

Publications   **161**

List of figures   **162**

List of tables   **173**

Introduction

Stratospheric ozone plays the essential role in shielding the surface of our planet from harmful ultraviolet (UV) radiation. By screening out solar radiation that has the potential to influence or destroy DNA information and cause mutations, ozone protects life on Earth. Furthermore, ozone absorbs and emits radiation in the infrared (IR) spectral range and, therefore, has an impact on the Earth's climate regarding the greenhouse effect. Accordingly, knowledge of the ozone distribution and the understanding of transport, formation and loss processes is of major importance and a challenge of atmospheric physics and chemistry.

Ozone abundance in different regions of the atmosphere is influenced by a variety of dynamical and photochemical processes. Although it had been known that anthropogenic emissions of nitrogen and halogen species reduce the stratospheric ozone amount (Crutzen, 1970; Molina and Rowland, 1974), the discovery of the Antarctic ozone hole (Farman et al., 1985) was quite unexpected. It demonstrated dramatically that mankind holds the ability of altering the natural equilibrium between formation and loss of ozone on a global scale. Chlorofluorocarbons (CFC), which were commonly used as cooling agents, were identified as the precursor substances of the ozone destroying species. Their production was first regulated and, later, stopped in the Montreal Protocol (1987) and its amendments.

Of particular interest are the exchange processes between the troposphere and the stratosphere, as emissions in general, and particularly anthropogenic emissions, are happening in the troposphere. The tropical upper troposphere and lower stratosphere is the region of the Earth's atmosphere, where tropospheric air masses are efficiently transported into the stratosphere and where the stratosphere, is supplied with halogen-bearing compounds (Fueglistaler et al., 2009). Thus, observations in the tropical UT/LS are of major interest regarding stratospheric ozone depletion.

As implied by a recent publication (Ravishankara et al., 2009), nitrogen oxides play an important role in the catalytic destruction of stratospheric ozone and will probably do so even more in future, as the precursor substance N_2O, which is also a potent greenhouse gas, remains presently unregulated by the Montreal Protocol. The Ozone Depleting Potential (ODP) of N_2O is strongly depending on the partitioning of NO_x and its reservoir species and therefore on photolytic lifetimes of both.

In the past decade balloon-borne measurements have evolved into a powerful tool for the investigation of photochemical processes relevant for stratospheric ozone chemistry. Among these tools spectroscopic techniques were developed to remotely detect a wide range of stratospheric trace gases via their spectral signatures which cover virtually all wavelengths ranging from the UV-B over the near-IR and mid-IR into the microwave (DOAS(Differential Optical Absorption Spectroscopy) (Ferlemann et al., 2000), SAOZ (Pommereau and Goutail, 1988), LPMA (Camy-Peyret et al., 1995), MIPAS-B (Oelhaf et al., 1991), MARK IV, TELIS (Birk et al.)). Due to the strong wavelength dependence of the atmospheric or solar emissivities, a variety of different instruments were designed to passively monitor the mid-IR to microwave atmospheric spectral emission or the atmospheric absorption in the UV to mid IR, either by direct sun observations (Ferlemann et al., 2000), or by limb scattered skylight spectrometry (Weidner, 2005; Bovensmann et al., 1999). In particular, the UV/vis spectral range supports the atmospheric monitoring of several important stratospheric radicals such as O_3, NO_2, NO_3, HONO, BrO, IO, OIO, OClO and CH_2O_2 on either satellite or balloon platforms.

While the satellite observations provide a nearly global coverage of the targeted species, balloon-borne UV/vis spectroscopic measurements of limb scattered skylight are powerful with respect to vertical and temporal resolution. Since balloon-borne observations are Lagrangian in nature, UV/vis limb measurements also support the monitoring of stratospheric radicals at changing illumination, and thus enable the study of their diurnal variation (McElroy, 1988; Roscoe and Pyle, 1987) in dependency of the solar zenith angle (SZA).

The measurement obtained for the present study are an extension of our previous balloon-borne UV/vis scattered skylight limb observations using the DOAS method (Weidner, 2005). While the previous studies mainly addressed balloon ascent measurements at fixed elevation angles, the field application was expanded in the framework of this thesis to include limb scanning observations at balloon float altitude.

These measurements were performed from various azimuth angle controlled balloon gondolas, such as LPMA/DOAS (Laboratoire de Physique Moléculaire et Applications and Differential Optical Absorption Spectroscopy), LPMA-IASI (Infrared Atmospheric Sounding Interferometer) and MIPAS-B (Michelson Interferometer for Passive Atmospheric Sounding-Balloon). The present thesis addressed the methods involved and the interpretation of the results with respect to stratospheric photochemistry.

Therefore a new algorithm is developed, that takes into account the temporal distance between the measurement and the state and is therefore well suited for the retrieval of the diurnal variation of UV/vis absorbing radicals. Since the forward model requires no chemical modeling as input, the retrieval method provides a tool for testing photochemical parameters. The method employs spectroscopic scanning observations in combination with radiative transfer modeling (RTM) and a mathematical inversion on a regularized time and height grid.

The presented study is outlined as follows. Chapter 1 highlights processes of atmospheric dynamics and photochemistry that are important for the interpretation of the measurements. Chapter 2 recapitulates the basics of radiative transfer in the Earth's atmosphere. Chapter 3 gives a description of the mini-DOAS instrument and its performance and introduces crucial parameters for radiative transfer modeling. Several steps are necessary to retrieve the diurnal variation of radicals from our spectroscopic measurements; a detailed description of the methods involved is given in chapter 4. Finally, results are presented in Chapter 5, where flight conditions are described for each balloon flight, followed by the presentation of measured quantities and retrieved concentrations. Comparison studies of our data with several other techniques are followed by studies concerning photochemistry.

Chapter 1

Atmospheric dynamics and photochemistry

The atmosphere serves as a transition zone between the Earth body and the space and provides the medium for life on the surface of the planet. There is a long tradition in human history in observing phenomena of the atmosphere, like the weather. New developments, such as balloons, rockets and satellites made it possible to investigate features of even higher atmospheric regions opening up a new field of research.

Atmospheric dynamics and photochemistry are the dominating processes in the determination of trace gas abundances in the upper troposphere and lower stratosphere.

Knowledge about these processes is provided in the following chapter, since they are essential for the understanding and interpretation of the trace gas measurements, presented in this work. Since our measurements provide a tool for the investigation of the upper troposphere and the lower and middle stratosphere, while the presented measurements are performed in the tropics, the focus is set to that particular region.

1.1 Vertical structure of the atmosphere

The atmosphere is the gaseous layer surrounding our planet, attached to it by gravity. The total atmospheric mass is $5.148 \cdot 10^{18}$ kg. Atmospheric pressure is a direct result of the total weight of the air above the point where the pressure is measured. Under normal conditions (i.e. SATP, Standard Ambient Temperature and Pressure: 25°C) at sea level, the pressure is $p_0 = 1013$ hPa := 1 atm, resulting in an air number density of $2.5 \cdot 10^{19}$ cm^{-3}. The barometric formula describes pressure as a function of altitude, assuming constant temperature t_0 throughout the atmosphere.

$$p(z) = p_0 \exp -\frac{z}{z_0} \tag{1.1}$$

with the scale height

$$z_0 = \frac{k \cdot T}{m \cdot g} \tag{1.2}$$

14 CHAPTER 1. ATMOSPHERIC DYNAMICS AND PHOTOCHEMISTRY

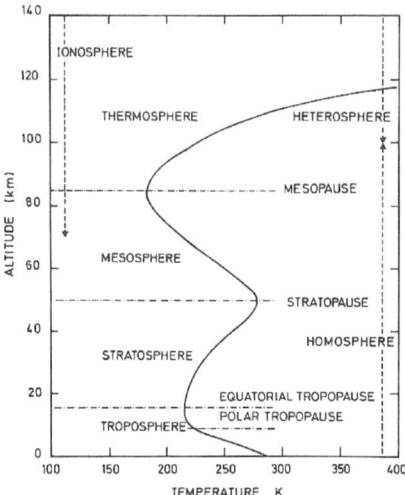

Figure 1.1: Vertical temperature profile of the Earth's atmosphere. Adopted from Brasseur and Solomon (1986).

In order to give a rough estimate of the atmospheric pressure distribution, z_0= 8 km is a good choice. This scale height z_0 implies halving of the pressure every 5.5 km. Of course the assumption of constant temperature is a bit audacious in view of Figure 1.1. Especially at colder temperatures, e.g. in the stratosphere, the pressure decreases faster than Equation 1.1 implies, leading to scale heights of only about 6 km. Figure 1.1 shows the vertical temperature profile of the atmosphere. On the basis of this profile the atmosphere can be divided into several layers with distinct boundaries, called pauses, where the temperature gradient changes its sign.

The lowermost layer, called troposphere, is the zone where weather phenomena and atmospheric turbulence are most marked. The troposphere contains about 75% of the total mass of the atmosphere and virtually all the water vapour and aerosols. The temperature profile in the troposphere is governed by adiabatic expansion and compression of airmasses during vertical transport. The driving force of this motion is the Sun. The solar radiation causes convection during the day, when the Earth's surface is heated up. While rising air parcels are cooled by expansion. Additional cooling occurs in the upper troposphere by radiative cooling in the infrared (IR) wavelength range mainly by water vapour, CO_2, CH_4 and O_3 (Goody and Yung, 1989, e.g.). These effects result in a lapse rate of 5-10 K/km and a temperature minimum at the top of the water vapor convective atmosphere called tropopause. The World Meteorological Organisation (WMO) defines the tropopause level as the lower boundary of a layer in which the vertical decrease in temperature is less than 2 K/km for a depth of at least 2 km. The height of the tropopause is linked to the height of the tropospheric water content and is therefore determined by the surface temperature. Furthermore, the temperature of the tropopause is colder the higher it is located. As a result, the tropical tropopause lies between 17 - 18 km with temperatures of around -80°C (the tropical

1.2. ATMOSPHERIC COMPOSITION

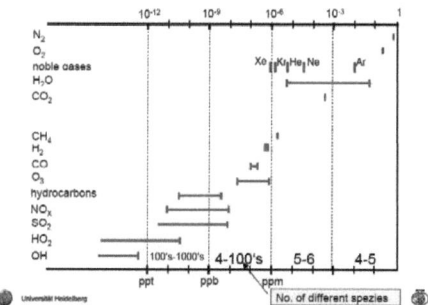

Figure 1.2: Gaseous constituents of the atmosphere, with their order of magnitude in abundance and the number of different species occurring in that particular range. Adopted from: http://www.iup.uni-heidelberg.de/institut/studium/lehre/Atmosphaerenphysik/.

tropopause layer (TTL) is discussed in more detail in section 1.5). Between 9 - 13 km, with temperatures of around -50°C at higher latitudes, with the lowest tropopause heights occurring in the polar winter atmosphere. The tropopause minimum acts as a barrier between the troposphere and stratosphere, because mixing and heat transport by convection can only occur when temperature decreases with height. The troposphere - with convection allowed - is turbulent and well mixed.

The conditions in the stratosphere are more or less vice versa. Here the temperature is increasing with altitude due to radiative cooling in the tropopause and heating in the upper stratosphere by the increasing absorption of solar radiation due to ozone and O_2. As a result convection is suppressed, making the stratosphere a stably stratified layer with very slow vertical exchanges. In contrast to the troposphere, the radiative budget is determined by absorption of solar radiation and emission of thermal IR radiation. The stratopause, which is given by a temperature maximum at around 50 km, separates the stratosphere from the mesosphere where temperatures decrease again. Above the mesopause at around 75 - 80 km the temperature is again strongly increasing due to the absorption of solar ultraviolet (UV) radiation, mainly by oxygen, up to values of 1200-1500 K. Therefore this layer is called thermosphere. The higher atmospheric layers (above 50 km) are also called ionosphere because of the occurrence of ions and free electrons. Finally, the heterosphere - at altitudes above 100 km - denotes the layer where the elements are layered according to their mass, because of the absence of mixing.

1.2 Atmospheric composition

The major atmospheric gaseous constituents, their abundances and the variety of different species in each range are shown in Figure 1.2. With about 78% and 21% respectively, N_2 and O_2 are the most abundant gases in the atmosphere. The remaining 1% of the atmospheric gases are known as trace gases because they are only present in small concentrations. The most abundant among them is the noble gas argon ($\approx$ 1% by volume). Noble gases, including neon, helium, krypton and xenon, are inert and do not engage in any chemical transformation within the atmosphere.

Despite their relative scarcity, the most important trace gases in the Earth's atmosphere are the greenhouse gases. Most abundant in the troposphere, these gases include carbon dioxide, methane, nitrous oxide, water vapour and ozone. According to their name, they are involved in the Earth's greenhouse effect, which keeps the planet naturally about $33°C$ warmer than it would be without an atmosphere. Apart from water vapor, the most abundant greenhouse gas is carbon dioxide. However, through emissions of greenhouse gases, mankind has increased i.e. the abundance of carbon dioxide in the past 50 years from 314 to 387 ppmv (Trends, 2009) and enhanced the natural greenhouse effect (which now lead to a warming climate on Earth).

Ozone behaves like a greenhouse gas in the troposphere and is an air pollutant with harmful effects on the respiratory system of humans and animals. In the stratosphere, where its abundance is most significant within the ozone layer, it absorbs incoming UV from the Sun, protecting life on Earth from its harmful effects. Stratospheric ozone chemistry is described in detail in section 1.7.

Other trace gases in the atmosphere arise from natural phenomena such as volcanic eruptions, lightning and forest fires. Gases from these sources include nitric oxides (NO_x), which play a major role for the abundance of ozone (see section 1.8). The production of NO_x by lightning and its further evolution are elaborated in section 1.5.1.

In addition to natural sources of nitric oxide and sulphur dioxide many man-made sources exist, including emissions from cars, agriculture and electricity generation through the burning of fossil fuels. During the 20th century other man-made processes emitted completely new trace gases into the atmosphere, for example the chlorofluorocarbons (CFCs, see section 1.7), which damage the stratospheric ozone layer.

Further important atmospheric constituents are the aerosols. They are solid or liquid particles dispersed in the air, and include dust, soot, sea salt crystals, spores, bacteria, viruses and a plethora of other microscopic particles. Collectively, they are often regarded as air pollution, but many of the aerosols are of natural origin. They are conventionally defined as those particles suspended in air having diameters in the region of 0.001 to 10 microns. They are formed by the dispersal of material at the surface (primary aerosols), or by reaction of gases in the atmosphere (secondary aerosols). Primary aerosols, which make 75% of all aerosols, can originate from volcanic dust, sea spray, organic materials from biomass burning, soot from combustion and mineral dust from wind-blown processes. The remaining 25% are called secondary aerosols, because the are generated by conversion of gaseous components to small particles through chemical processes. Although making up only 1 part in a billion of the mass of the atmosphere, aerosols have the potential to significantly influence the amount of sunlight that reaches the Earth's surface, and therefore the Earth's climate. Their influence on the radiative transfer also effects our measurements in a way, that on one hand the measurements provide information on aerosols and, on the other hand, if not accounted for they lead to errors in the retrieval of trace gas profiles.

The variety of shapes, sizes and densities leads to a variety in the optical properties of aerosols, making them difficult to handle in radiative transfer, which they decisively influence. Although the abundance of aerosols varies over short time scales, for example, after a volcanic eruption, on the long term the atmosphere is naturally cleansed through mixing processes and rainfall. Cleansing is never complete however, and there exists a natural background level of aerosols in the atmosphere. Their number density N decreases with height, as shown in Figure 1.3, but at an altitude around 20 km, there is a layer with enhanced aerosol content, called Junge layer.

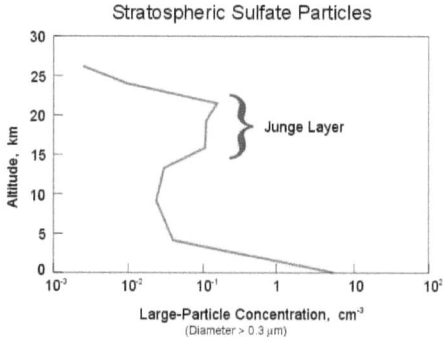

Figure 1.3: Stratospheric aerosol layer (Junge layer). Concentration of large particles (diameter ≥ 0.3μm). Adopted from Chagnon and Junge (1961).

1.3 Dynamics in the troposphere

The troposphere can be subdivided into several layers according to their dynamics. The lowermost millimeters form the molecular viscous layer, named by the dominating molecular viscosity while above turbulent diffusion is the driving force for dynamics. In the Prandtl layer, from 20 - 200 m, surface friction is most relevant. The transition to the free troposphere above around 1000 m, where dynamics are driven by global circulation patterns occurs in the Ekman layer. Here the wind direction changes steadily from the ground wind direction to the direction of the quasi geostrophic winds of the free troposphere. The combination of all these layers forms the steadily mixed planetary boundary layer (PBL), where most of the pollution occurs.

Figure 1.4 illustrates the global circulation patterns in the free troposphere, which are briefly discussed in the following. Strong differences in solar radiation between the tropics and higher latitudes are the driving force of the global circulation. In the tropics and up to around 30°-35° northern and southern latitude the trade winds are the main wind system. They blow at the ground from northeast on the northern and from southeast on the southern hemisphere. When they meet in the equatorial region, they form the Inner Tropical Convergence zone (ITC). The trade winds are a result of the uplift of hot humid air in the zone of the strongest solar radiation, resulting in a low pressure belt and an inflow of air from further south and north. The Coriolis force is distracting the air to the west. This causes the air pressure to be rather low in the ITC and to increase towards higher latitudes where the air descends again. This circulation pattern is referred to as the Hadley cell. Around 30°-35° N and S, there is the subtropical high pressure belt, which is characterized by high pressure, regularly calm winds and a vertical wind component directed downwards. Further poleward, from around 35°-70° N and S, the zone of west wind drift attaches with typically westerly winds which are not as uniform as in the tropics but disturbed by cyclones, anti-cyclones and waves of different wavelengths. The west wind drift are thermal winds caused by the temperature gradient between (sub-)tropical and higher latitudes. Cold air masses from high latitudes and warmer air masses from moderate latitudes meet at the polar front. This

18   CHAPTER 1. ATMOSPHERIC DYNAMICS AND PHOTOCHEMISTRY

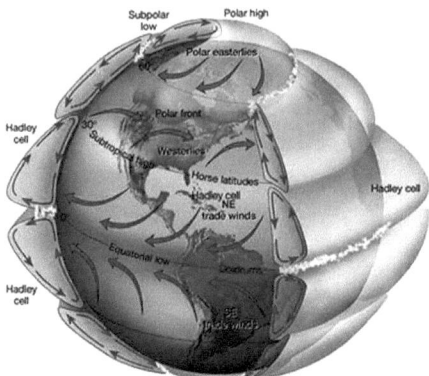

Figure 1.4: Global circulation. Adopted from http://rst.gsfc.nasa.gov/Sect14/Sect14$_1$c.html.

region is characterized by low pressure, increasing towards the poles. At high polar latitudes there are circumpolar east winds at lower altitudes caused by downward winds that are deflected eastwards by the Coriolis force.

1.4 Dynamics in the stratosphere

Figure 1.5 illustrates the stratospheric dynamics by a schematic of a meridional cross section of the Earth's atmosphere. Thermal gradients between the tropics and high-latitudes lead to strong winds in zonal direction, with the sign depending on season. On the summer hemisphere the gradient is vice versa compared to the troposphere, since the summer polar stratosphere is heated by absorption of solar radiation. The gradient from the summer pole to the equator causes strong thermal easterlies, winds in east-west direction. The winter polar stratosphere is colder than its tropical counterpart due to the lack of insulation, which results in strong westerlies surrounding the winter pole, called the polar vortex. Due to a circumpolar land mass, surrounded by the ocean, the antarctic polar vortex generally evolves unperturbed and isolates polar from mid-latitudinal air. This obviation of meridional mixing of air masses induces conditions for photochemical processes that lead to the formation of the ozone hole. The northern hemisphere is more structured with oceans and mountains, disturbing the stratospheric circulation and leading to a stronger meridional mixing and warming of the winter polar region.
Because of those geographic differences the ozone hole over the Antarctic is more severe than in the Arctic.
The Brewer-Dobson circulation basically describes the transport from the tropics to the poles, which is governed by a large scale diabatic circulation. Main activator is the dissipation of tropospheric waves propagating into the stratosphere, resulting in rising motion in the tropics, poleward flow at mid-latitudes and sinking motion in the polar regions. The amplitude of such waves, initiated for example by mountains, increases exponentially with altitude when propagating into the stratosphere due to the decrease in air density. The energy and momentum of those waves, dissipating by the ambient atmosphere, to-

1.4. DYNAMICS IN THE STRATOSPHERE

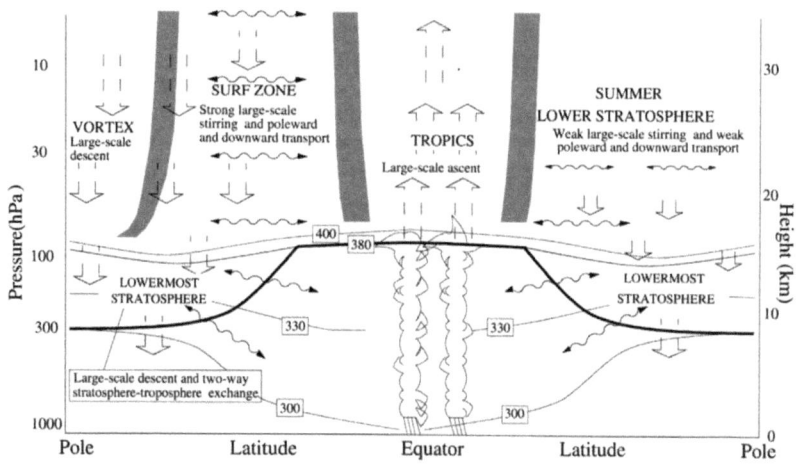

Figure 1.5: General circulation patterns in the atmosphere. For details see text. Adopted from WMO (1999).

gether with the Coriolis force result in a net poleward motion. Since this initial activity occurs mainly at mid-latitudes the mechanism is often referred to as the "extra tropical pump" (Holton et al., 1995). The meridional transport is quite fast, as indicated by the "surf zone" in Figure 1.5.

The Brewer-Dobson circulation shows a strong seasonal cycle, because the propagation of tropospheric waves into the stratosphere is most effective in winter. Topographic differences lead to a stronger magnitude and frequency of those waves in the northern hemisphere, making the meridional flow in the northern hemisphere more pronounced than in its southern counterpart. Important barriers for this flow are indicated as black shaded regions in Figure 1.5. The edge of the polar vortex forms the so called "polar transport barrier" by conservation of potential vorticity. Between 20° N/S and 30° N/S latitude the subtropical transport barrier separates the horizontally well mixed surf zone from the tropical region where transport in vertical direction is dominating. This tropical region, often referred to as "tropical pipe" (Plumb, 1996) is isolated from mid-latitudes, and trace species budgets there are balances between mean upwelling and local chemical sources and sinks.

The Quasi Biannual Oscillation (QBO) is an oscillation of the east-west wind in the tropical stratosphere. The QBO effect occurs throughout the tropics, but it is most often shown as a change in the direction of the stratospheric zonal wind at Singapore. Figure 1.6 shows these zonal winds at Teresina (5°S, 317°E) from 2001 - 2008 between up to 48 km altitude. The wind direction over the tropical stratosphere changes sign (direction) about every year. However, because the QBO is due to the internal dynamics of tropical waves rather than the annual change of seasons cycle, the period of this wind oscillation is highly variable with periods ranging from 22 to 34 months. Hence the name, quasi-biennial oscillation, reflects the variable period of this phenomenon.

20   CHAPTER 1. ATMOSPHERIC DYNAMICS AND PHOTOCHEMISTRY

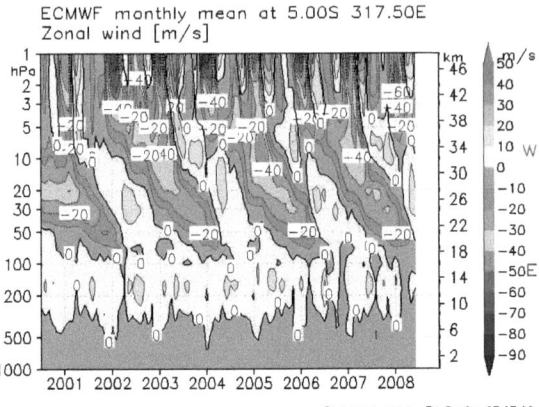

Figure 1.6: ECMWF calculated time-height section of the monthly mean of zonal wind [m/s] at Teresina, northern Brazil, at 12 UT, in 2005 (courtesy of Katja Grunow, FU Berlin).

The QBO is observed between 20-35 km, where the easterly winds are generally stronger than the westerly winds. Westerly wind regimes descend faster in time and persist longer at lower levels than easterly wind regimes. Below 15 km, there is little evidence of the QBO, while above 35 km, the QBO coexists with another regular oscillation of the mesosphere known as the semiannual oscillation.

1.5 Stratospheric - tropospheric exchange: The tropical tropopause layer

Of particular interest are the exchange processes between the troposphere and stratosphere, as emissions in general, and particularly anthropogenic emissions, are happening in the troposphere. The impact on the stratosphere depends on exchange processes between the troposphere and the stratosphere. The low stratospheric water content suggests that the main entry to the stratosphere occurs via the very cold tropical tropopause, which acts as a cold-trap for water. Observations of temperature, winds, and atmospheric trace gases suggest that the transition from the troposphere to the stratosphere occurs in an extended layer, rather than at a sharp "tropopause". In the tropics, this layer may vertically extend over several kilometers and is often referred to as the "tropical tropopause layer" or "tropical transition layer"(TTL). In the TTL we can find both, tropospheric and stratospheric characteristics, but, depending on the quantity, the extent of the layer might be shifted up or down. Figure 1.7 shows several quantities related to troposphere or stratosphere and their extent in the tropopause, which is assumed to range from 150 hPa (14 km) to 70 hPa (18.5 km). Laterally, the TTL is bounded by the position of the subtropical jets. The TTL is of interest not only because of being the interface between two very different dynamical regimes, but also because it acts as a "gate to the stratosphere" for atmospheric tracers such as water vapor and so-called very short lived (VSL) substances, which both play an important role in stratospheric chemistry (Fueglistaler et al., 2009).

1.5. STRATOSPHERIC - TROPOSPHERIC EXCHANGE: THE TROPICAL TROPOPAUSE LAYER

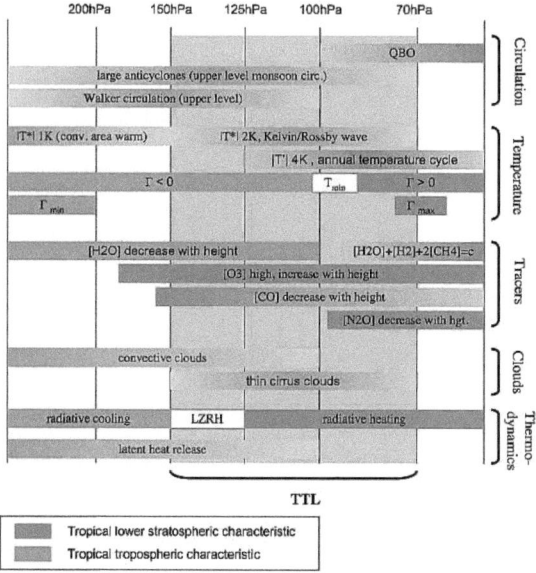

Figure 1.7: Summary of tropospheric/stratospheric characteristics and transitions thereof (symbolically shown as fade out of colored pattern).Θ, temperature lapse rate; Tmin, temperature minimum of profile; |T*|, amplitude of quasi-stationary zonal temperature anomaly; |T|, amplitude of tropical mean temperature seasonal cycle; QBO, quasi-biennial oscillation. Adopted from Fueglistaler et al. (2009).

Figure 1.8 shows a schematic of cloud processes and transport on the left panel. On the right panel the zonal mean circulation is outlined, where arrows indicate circulation, the black dashed line is the clear-sky level of zero net radiative heating (LZRH) and the black solid lines show isentropes in K, based on European Centre for Medium Range Weather Forecasts 40-year reanalysis (ERA-40). The letters indicate the following processes:
a) Deep convection with the main outflow at around 200 hPa and a rapid decay with height in the TTL serves as a fast vertical transport of tracers from the boundary layer into the TTL.
b) Radiative cooling in the troposphere leads to subsidence of air.
c) Subtropical jets limit quasi-isentropic exchange between the troposphere and the stratosphere (transport barrier).
d) Radiative heating balances forced diabatic ascent.
e) Rapid meridional transport of tracers and mixing occurs in the TTL.
f) The edge of the "tropical pipe" leads to a relative isolation of the tropical stratosphere and a stirring of air flow over extra tropics, which is referred to as the "surf zone".
g) A usual deep convective cloud ending at the boundary to the TTL.
h) A deep convective cloud with the convective core overshooting its level of neutral buoyancy.
i) Ubiquitous optically (and geometrically) thin, horizontally extensive cirrus clouds, often formed in

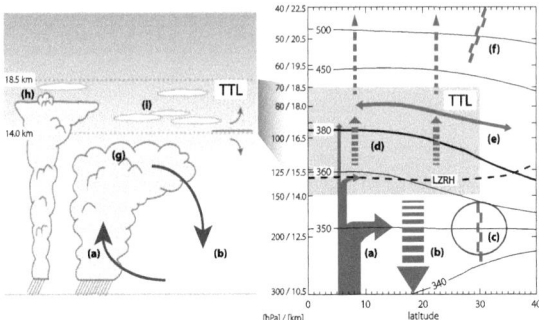

Figure 1.8: Schematic of cloud processes and transport (left) and of zonal mean circulation (right) belonging to the TTL. Arrows indicate circulation, the black dashed line is the clear-sky level of zero net radiative heating (LZRH) and the black solid lines show isentropes in K, based on European Centre for Medium Range Weather Forecasts 40-year reanalysis (ERA-40). The relations between height, pressure and potential temperature are based on tropical annual mean temperature fields, with height values rounded to the nearest 0.5 km. The letters are explained in the text. Adopted from Fueglistaler et al. (2009).

situ.

1.5.1 Convection and lightning

Electrical discharges as they are occurring in thunderstorms can heat air masses up to several 1000 K, where molecular bonds of O_2, N_2 and H_2 break and a temperature dependent equilibrium between atomic and molecular species builts up. As a result NO is formed through the following reactions (Ya. B. Zeldovich, 1966).

$$O + N_2 \longrightarrow NO + N \qquad (R1.1)$$

$$N + O_2 \longrightarrow NO + O \qquad (R1.2)$$

The production of NO_2 in the heated airmass requires 2 molecules of NO, which explains the up to three orders of magnitude lower fraction of NO_2 in the temperature range of 3000 to 4000 K.

$$2 NO + O_2 \longrightarrow 2 NO_2 \qquad (R1.3)$$

When combined with convection, NO_x from near surface lightning is transported together with other trace gases to the upper troposphere and sometimes even the lower stratosphere (Huntrieser et al., 2008). Via moist convection, air from near Earth's surface is rapidly transported upward and detrained into the UT. In this process nitric acid (highly soluble) is efficiently scavenged while NO_x (insoluble) remains. NO_x levels are increased by concurrent lightning NO production, resulting in high NO_x/HNO_3 ratios in the convective outflow region of a thunderstorm cloud system. After detrainment into the UT, NO_x is converted to HNO_3 by OH during the day and through reaction with NO_3 at night, followed by hydrolysis

1.5. STRATOSPHERIC - TROPOSPHERIC EXCHANGE: THE TROPICAL TROPOPAUSE LAYER

of the N_2O_5 product. The chemical evolution of the NO_x/HNO_3 ratio provides a unique indicator of the length of time that a sampled air mass has spent in the UT after convection (Bertram et al., 2007).

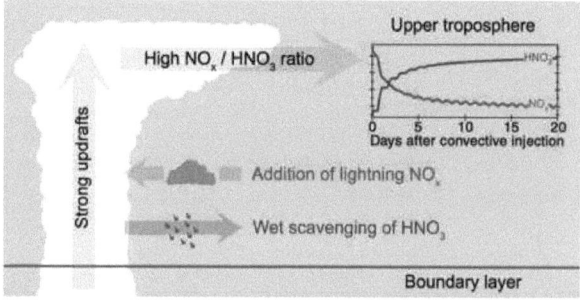

Figure 1.9: In moist convection, air from near Earth's surface is rapidly transported upward and detrained into the UT. In this process, nitric acid (highly soluble) is efficiently scavenged while NO_x (insoluble) remains. NO_x is elevated by concurrent lightning NO production, resulting in high NO_x/HNO_3 ratios in the convective outflow region. After detrainment into the UT, NO_x is converted to HNO_3 by OH during the day and through reaction with NO_3, followed by hydrolysis of the N_2O_5 product, at night. The chemical evolution of the NO_x/HNO_3 ratio provides a unique indicator of the length of time that a sampled air mass has been in the UT after convection. Adopted from Bertram et al. (2007).

1.6 Chemistry of HONO in the troposphere

As unexpected detection of HONO in the upper troposphere appears to be associated to lightning, the role and possible formation of HONO is discussed in the following.
HONO (Nitrous acid) is a source of the most important daytime radical, OH (the hydroxyl radical) (e.g. Alicke et al. (2003)). The OH radical is one of the key species in photochemical cycles responsible for tropospheric ozone formation, which can lead to the so called "photochemical smog" in polluted regions. OH is released from HONO by photolysis (e.g. Cox (1974))

$$HONO + h\nu \longrightarrow OH + NO \qquad (R1.4)$$

HONO can be formed in the troposphere by heterogeneous, homogeneous or photochemical processes (Stutz et al., 2002). The following figure 1.10 displays a plumb diagram of possible reactions leading to HONO formation and loss in the troposphere.

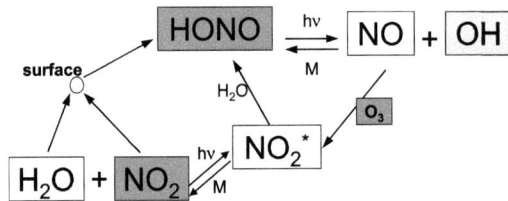

Figure 1.10: Possible reactions leading to HONO formation and loss in the upper troposphere

The only important gas-phase reaction forming HONO is the reaction of NO and OH (Pagsberg et al., 1997),

$$NO + OH \longrightarrow HONO \qquad (R1.5)$$

together with the back reaction, the photolysis of HONO (reaction R1.4) builds the photochemical steady state. The reaction of exited NO_2 (marked with a star) and water

$$NO_2* + H_2O \longrightarrow HONO \qquad (R1.6)$$

is negligible, under tropospheric conditions, because the deactivation by hitting air molecules

$$NO_2* + M \longrightarrow NO_2 + M \qquad (R1.7)$$

is much faster than the reaction with water. There are additional heterogeneous reactions of water and nitrogen dioxide on surfaces such as the ground, walls or aerosols. Observations in the laboratory indicate that nitrous acid is formed heterogeneously in the presence of nitrogen oxides and water. The two following mechanisms have been proposed (Stutz et al., 2002).

$$NO + NO_2 + H_2O \longrightarrow 2HONO \quad (R1.8)$$
$$2NO_2 + H_2O \longrightarrow HONO + HNO_3 \quad (R1.9)$$

1.7 Photochemistry of ozone in the stratosphere

The chemistry in the stratosphere is not only influenced by dynamics and transport processes, but also driven by the solar radiation, which is called photochemistry.
Ozone is of major importance for the chemical and radiative budget in the stratosphere and thereby also for life on Earth (its low abundance it retends 95 to 99% of the short wave radiation in 20 to 50 km).
Reaction cycles involving oxygen, hydrogen, nitrogen and halogen containing species govern the formation and destruction of stratospheric ozone.
In 1930 Sidney Chapman (Chapman, 1930) published the first simple theory on stratospheric ozone, involving only oxygen species. Molecular oxygen (O_2) is photolysed by ultraviolet radiation with wavelengths below 242 nm. The oxygen atoms react with molecular oxygen to ozone via a three-body reaction:

$$O_2 + h\nu \longrightarrow 2O \quad (R1.10)$$
$$O + O_2 + M \longrightarrow O_3 + M \quad (R1.11)$$

$$(R1.12)$$

Ozone loss happens through photolysis by UV radiation with wavelength below 310 nm followed by the reaction with another oxygen atom or ozone:

$$O_3 + h\nu \longrightarrow O_2 + O(^3P) \quad (R1.13)$$
$$O_3 + h\nu \longrightarrow O_2 + O(^1D) \quad (R1.14)$$
$$O(^1D) + M \longrightarrow O(^3P) + M \quad (R1.15)$$
$$O + O + M \longrightarrow O_2 + M \quad (R1.16)$$
$$O(^3P) + O_3 \longrightarrow 2O_2. \quad (R1.17)$$

Due to the strong attenuation of solar UV radiation, photolysis of molecular oxygen and thus O_3 production occurs mainly in the upper stratosphere. Together with photo dissociation of O_3 (reaction R1.13 and R1.14) an equilibrium builds up. Within this equilibrium, forward and backward reaction of atomic oxygen to O_3 happens in the order of seconds (R1.10 and R1.11). The rapid transformation of one species to the other allows the definition of families, where the lifetime of the whole family is rather long. The O_x family is defined as the sum of odd-oxygen like O_3 and $O(^1D)$.
Since the ozone concentration predicted by the Chapman-Cycle (R1.10-R1.17) was too large and the maximum of the layer too high, it seemed obvious that additional O_3 loss processes must exist. Bates

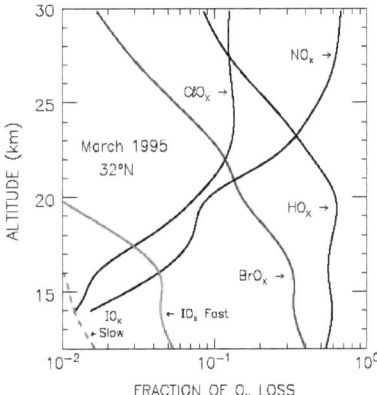

Figure 1.11: Fractional contribution to odd oxygen (O_x) loss by catalytic cycles involving nitrogen (NO_x), hydrogen (HO_x), chlorine (ClO_x), bromine (BrO_x), and iodine (IO_x), calculated for March 1995, 32° N, using JPL-2002 kinetics (Sander et al., 2003). Br_y was derived from the breakdown of CH_3Br and halons plus additional 5 ppt representing Br_y^{VSLS}. The ClO_x curve represents loss from the ClO+O and $ClO+HO_2$ cycles, plus other minor cycles that involve ClO, but not BrO or IO. The BrO_x curve represents loss from the BrO+ClO and $BrO+HO_2$ cycles, plus other minor cycles that involve BrO, but not IO. Adopted from WMO (2006).

and Nicolet (1950) first suggested ozone destruction via catalytic cycles,

$$X + O_3 \longrightarrow XO + O_2 \qquad (R1.18)$$
$$O + XO \longrightarrow X + O_2 \qquad (R1.19)$$
$$\text{net}: \; O + O_3 \longrightarrow 2\,O_2 \,. \qquad (R1.20)$$

where odd-oxygen is transformed into its reservoir XO by a catalyst X, which can be substituted by the radicals OH (Bates and Nicolet, 1950), NO (Crutzen, 1970; Johnston, 1971), Cl (Molina and Rowland, 1974), Br (Wofsy et al., 1975) and possibly I (Solomon et al., 1994). All these species react faster with ozone than atomic oxygen. The fractional contribution of the proposed catalytic cycles with respect to O_3 destruction is shown in Figure 1.11 and depends on the number of cycles completed before the catalyst X is lost in some chain termination reaction. There are additional cycles involving the $HO_x = H + OH + HO_2$ catalyst, e.g. the following, where O_3 is destroyed without the abundance of an O atom:

$$OH + O_3 \longrightarrow HO_2 + O_2 \qquad (R1.21)$$
$$HO_2 + O_3 \longrightarrow OH + 2\,O_2 \qquad (R1.22)$$
$$\text{net}: \qquad 2\,O_3 \longrightarrow 3\,O_2 \,. \qquad (R1.23)$$

1.8. PHOTOCHEMISTRY OF NITROGEN IN THE STRATOSPHERE

This cycle is most important at lower altitudes, where less atomic oxygen is available. Whereas the following cycle, involving atomic oxygen, dominates in the upper stratosphere:

$$OH + O \longrightarrow O_2 + H \quad \text{(R1.24)}$$
$$H + O_2 + M \longrightarrow HO_2 + M \quad \text{(R1.25)}$$
$$HO_2 + O \longrightarrow OH + O_2 \quad \text{(R1.26)}$$
$$\text{net:} \quad 2O \longrightarrow O_2. \quad \text{(R1.27)}$$

The combination of different families of catalysts leads to further catalytic cycles of O_3 destruction:

$$X + O_3 \longrightarrow XO + O_2 \quad \text{(R1.28)}$$
$$Y + O_3 \longrightarrow YO + O_2 \quad \text{(R1.29)}$$
$$XO + YO \longrightarrow X + Y + O_2 \quad \text{(R1.30)}$$
$$\text{net:} \quad 2O_3 \longrightarrow 3O_2, \quad \text{(R1.31)}$$

where possible candidates are: $X = OH$ and $Y = Cl$, $X = OH$ and $Y = Br$, $X = Br$ and $Y = Cl$.

In order to assess the relative importance of the various catalytic cycles regarding odd oxygen loss, Figure 1.11 shows their modeled fractional contributions for a mid-latitudinal station (32°N) in March 1995. From 12 to 25 km altitude the HO_x and BrO_x $(= Br + BrO)$ catalytic cycles are most important, while between 25 km and 40 km the O_3 destruction is dominated by NO_x. Since O_3 loss due to the coupled ClO-BrO cycle is completely assigned to BrO_x, the importance of BrO_x compared to ClO_x $(= Cl + ClO + 2 Cl_2O_2)$ in the lower stratosphere is somewhat disproportionate. In this calculation IO_x $(= I + IO)$ is only of minor importance.

1.8 Photochemistry of nitrogen in the stratosphere

The presence of nitrogen oxides in the stratosphere is mainly due to the oxidation of nitrous oxide (N_2O) and, to a lesser extend, the ionization of molecular nitrogen (N_2) by solar and galactic high energy particles.

$$N_2O + h\nu \longrightarrow N_2 + O(^1D) \quad \text{(R1.32)}$$
$$N_2O + O(^1D) \longrightarrow 2NO \quad \text{(R1.33)}$$
$$\longrightarrow N_2 + O_2. \quad \text{(R1.34)}$$

An anthropogenic source is the combustion in aircraft engines directly in the stratosphere.
Nitrogen oxides, meaning NO and NO_2 are referred to as the NO_x family, since they are in photochemical balance during daytime due to fast conversion mechanisms.

$$NO + O_3 \longrightarrow NO_2 + O_2 \quad \text{(R1.35)}$$
$$NO_2 + O \longrightarrow NO + O_2. \quad \text{(R1.36)}$$
$$NO_2 + h\nu \longrightarrow NO + O. \quad \text{(R1.37)}$$

NO reacts with O_3 forming NO_2, which is transformed back to NO either by photolysis or by reaction with atomic oxygen. The ratio of NO/NO_2 is called Leighton ratio and is almost constant and equal to one for altitudes below 40 km during daytime. The catalytic cycle converting NO/NO_2 provides a major loss process for odd oxygen in the stratosphere. The limiting process of this conversion cycle is the conversion into less reactive nitrogen compounds, like the reaction of NO_2 with O_3 to NO_3.

$$NO_2 + O_3 \longrightarrow NO_3 + O_2, \tag{R1.38}$$

which is rapidly photolysed during day:

$$NO_3 + h\nu \longrightarrow NO_2 + O. \tag{R1.39}$$

But in the night photolysis is switched off and, through the following reaction, significant amounts of N_2O_5 are built up until sunrise

$$NO_2 + NO_3 + M \longrightarrow N_2O_5 + M, \tag{R1.40}$$

with unimolecular decay being the only backreaction:

$$N_2O_5 + M \longrightarrow NO_2 + NO_3 + M. \tag{R1.41}$$

With the first beam of sunlight, photolysis starts again and transforms N_2O_5 at moderate rate back to the educts of its formation reaction,

$$N_2O_5 + h\nu \longrightarrow NO_2 + NO_3. \tag{R1.42}$$

The collisional decomposition of N_2O_5, is slow an therefore of minor importance in the stratosphere. The conversion of nitrogen oxides into nitric acid (HNO_3), which is a more stable reservoir compared to N_2O_5, occurs either through the reaction between NO_2 and OH

$$NO_2 + OH + M \longrightarrow HNO_3 + M \tag{R1.43}$$

$$\tag{R1.44}$$

or through heterogeneous hydrolysis of N_2O_5 on the surface of aerosol particles.

$$N_2O_5(g) + H_2O(s) \longrightarrow 2\,HNO_3(s), \tag{R1.45}$$

where (g) and (s) indicate gas and condensed phase, respectively.
Further long lived species are $ClONO_2$ and $BrONO_2$, which simultaneously act as reservoirs for ClO_x, BrO_x and NO_x radicals and thus buffer catalytic O_3 loss

$$NO_2 + ClO + M \longrightarrow ClONO_2 + M \tag{R1.46}$$
$$NO_2 + BrO + M \longrightarrow BrONO_2 + M,. \tag{R1.47}$$

The catalysts can be released from their reservoirs by photolysis as in Reaction R1.42.

$$HNO_3 + h\nu \longrightarrow OH + NO_2 \tag{R1.48}$$
$$ClONO_2 + h\nu \longrightarrow ClO + NO_2 \tag{R1.49}$$
$$BrONO_2 + h\nu \longrightarrow BrO + NO_2. \tag{R1.50}$$

1.8. PHOTOCHEMISTRY OF NITROGEN IN THE STRATOSPHERE

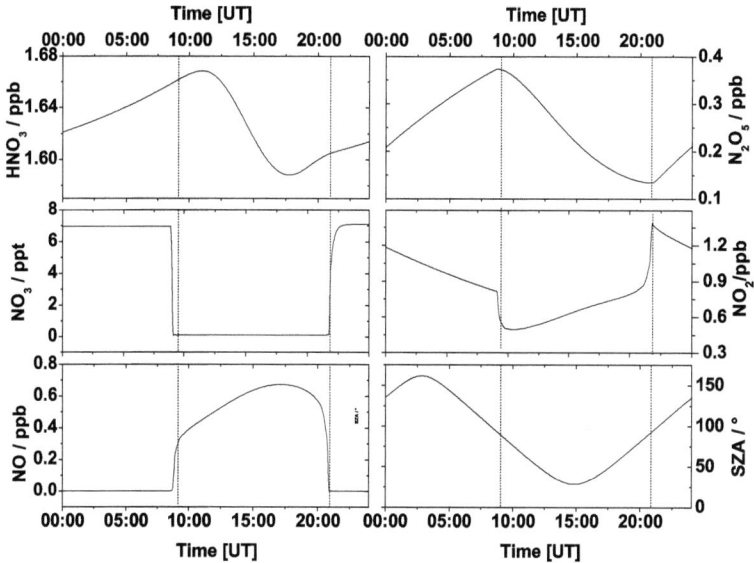

Figure 1.12: Diurnal variation of stratospheric nitrogen species. Temporal evolution of HNO$_3$ (upper left panel), NO$_3$ (middle left panel), NO (lower left panel), N$_2$O$_5$ (upper right panel), NO$_2$ (middle right panel) and SZA (lower right panel). The data are taken from a run of the LABMOS model of stratospheric chemistry on the 615 K potential temperature surface ($\approx$ 25 km). SZA = 90° is indicated by dotted vertical lines.

or in the case of HNO$_3$ via reaction with OH

$$\mathrm{HNO_3 + OH \longrightarrow H_2O + NO_3}. \tag{R1.51}$$

Hence, NO$_x$ plays an ambiguous role as a catalyst for O$_3$ loss on the one hand, and as a scavenger for ozone depleting radicals on the other and is therefore crucial for stratospheric ozone chemistry.

The output of a model run from the 1-D chemical model LABMOS shows the diurnal variation of some nitrogen species in Figure 1.12.

1.9 Photochemistry of halogens in the stratosphere

The presence of halogens (chlorine, fluorine, bromine and perhaps iodine) in the stratosphere is a consequence of halocarbons transported upward from the troposphere. They are released at the Earth's surface as a result of natural or anthropogenic processes. In the stratosphere the halocarbons are broken by photolysis and release halogen atoms, which react rapidly with ozone to form FO, ClO, BrO and IO. Halogens are of great importance in stratospheric ozone destruction, in particular under ozone hole conditions. The release of man-made halogen substances has disturbed the natural equilibrium between formation and destruction of ozone. Chemical partitioning processes play a major role in halogen-catalyzed ozone destruction, deciding whether a certain halogen is available to take place in an catalytic cycle, or bound in a reservoir. The reservoir compound hydrogen fluoride, HF, is so stable that fluorocarbons have relatively no known impact on ozone. The relative importance of the ClO_x, BrO_x and IO_x catalysts for ozone loss depends on their lifetime and distribution of their source gases, and on the efficiency of the respective catalytic cycles (see Figure 1.11).

Chlorine

The most prominent source of ClO_x in the stratosphere are the chlorofluorocarbons (CFCs). Since, in the troposphere they are chemically inert, photostable and with low water solubility, their lifetime is on the order of years to centuries resulting in an almost uniform distribution in the troposphere and effective transport to the stratosphere. The only natural source of stratospheric chlorine is methyl chloride (CH_3Cl), which contributes about 16.2% to the total budget (WMO, 2006). After photolysis and reaction with ozone, ClO_x undergoes rapid cyclic transformations between its constituents according to

$$Cl + O_3 \longrightarrow ClO + O_2 \qquad (R1.52)$$
$$ClO + O \longrightarrow Cl + O_2 \qquad (R1.53)$$
$$ClO + NO \longrightarrow Cl + NO_2 \qquad (R1.54)$$
$$ClO + OH \longrightarrow Cl + HO_2, \qquad (R1.55)$$

resulting in loss of odd oxygen. A great success in limiting the anthropogenic production of CFCs was the Montreal Protocol (1987) and its amendments (London (1990), Copenhagen (1992), Vienna (1995), Montreal (1997), Bejing (1999)), which limit the production of CFCs. They were partly replaced by hydrochlorofluorocarbons (HCFCs), which are less stable in the troposphere and thus have shorter lifetimes. Figure 1.13 shows a record of the long-term trend of active chlorine in the middle stratosphere. The trend shows a rapid rise of 58% from 1982 to a broad maximum in 1994–1997 and a substantial decline of 1.5% /year from the maximum through late 2004 (Solomon et al., 2006). About 1/3 of the decline may be due to changes in methane, which reacts with the ClO_x catalyst to form temporary reservoir species as with with NO_2 and HO_2

$$ClO + NO_2 + M \longrightarrow ClONO_2 + M \qquad (R1.56)$$
$$ClO + HO_2 \longrightarrow HOCl + O_2 \qquad (R1.57)$$
$$Cl + HO_2 \longrightarrow HCl + O_2 \qquad (R1.58)$$
$$Cl + CH_4 \longrightarrow HCl + CH_3. \qquad (R1.59)$$

1.9. PHOTOCHEMISTRY OF HALOGENS IN THE STRATOSPHERE

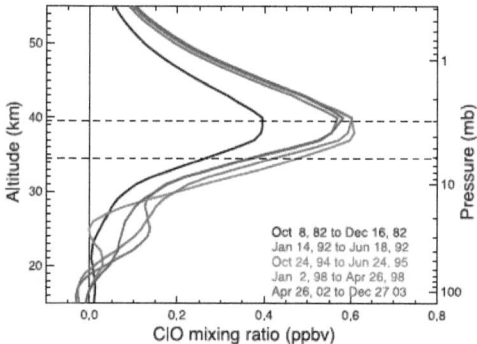

Figure 1.13: Altitude profiles of ClO over Hawaii (20° N) retrieved from spectra averaged over the indicated periods, illustrating the rapid rise in ClO from the early 1980s to the mid 1990s, the leveling off in the mid 1990s and a substantial decline by 2003 to about the 1992 level. Note that the 2003 and 1992 profiles almost overlap. Adopted from Solomon et al. (2006).

The active species can be regenerated by photolysis

$$ClONO_2 + h\nu \longrightarrow Cl + NO_3 \quad (R1.60)$$
$$\longrightarrow ClO + NO_2 \quad (R1.61)$$
$$HOCl + h\nu \longrightarrow OH + Cl. \quad (R1.62)$$

The contribution of VSL chlorine species to the total stratospheric chlorine budget is estimated to be less than 5% of the total chlorine load and hence their effect is considered to be small (WMO, 2006).

Bromine

Source gases for the BrO_x catalyst are originating from anthropogenic and natural sources in comparable amounts (see Figure 1.14). The major source is methyl bromide, which provides 30% of the total bromine content (WMO, 2006) and is released by natural (biomass burning, oceans) and anthropogenic (agricultural fumigants, gasoline additives, etc.) processes with a lifetime of 0.7 years. Halogenated hydrocarbons (halons) produced to extinguish fires, contribute about 30% - 40% to the global bromine budget. Their global abundance is still increasing by 0.1 ppt/year due to the ongoing use of halon stocks and halon production in developing countries (Montzka et al., 2003). The tropospheric bromine loading from

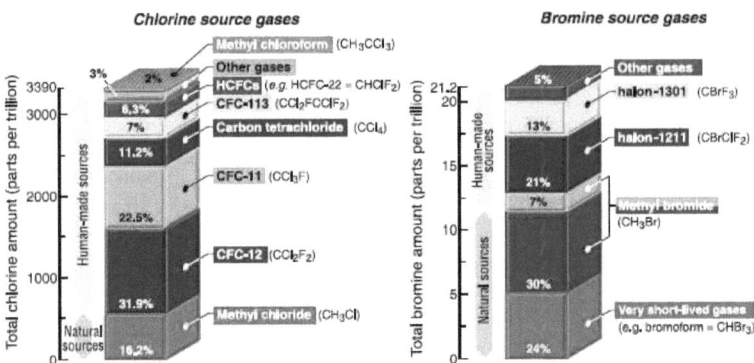

Figure 1.14: Overview about the chlorine (left) and bromine (right) primary source gases in 2004. Adopted from WMO (2006).

halons and methyl bromide peaked around 1998 at 16 - 17 ppt and started to decline afterwards by a mean annual rate of (0.25 ± 0.09) ppt/year (Montzka et al., 2003). The Latest balloon-borne measurements of BrO by Dorf (2005) indicate a total inorganic bromine burden of (19 ± 3) ppt and, consistent with the tropospheric trend, the increase of stratospheric total bromine is slowing down or leveling off. Figure 1.15 illustrates the trend of the tropospheric and stratospheric bromine throughout the past 20 years. There is a discrepancy between total inorganic stratospheric bromine and the sum of long-lived tropospheric organic precursors, which is commonly explained by a contribution of very short-lived brominated source gases or a direct upward transport of inorganic bromine across the tropopause (WMO, 2006; Salawitch et al., 2005; Dorf, 2005).

Coastal areas in the tropics and subtropics show the highest concentrations of VSL bromocarbons, indicating oceanic emissions, which account for 90% to 95% of brominated and iodinated VSL species to the atmosphere. Anthropogenic emissions of most VSL substances contribute only little to the global budget although being locally significant. Due to efficient vertical transport, the above mentioned regions provide a suitable candidate for the missing source of bromine in the stratosphere. It is subject of ongoing research what fraction of bromine released as VSLS enters the stratosphere as organic source gas (SG) or inorganic product gas (PG) produced from the decomposition of VSL SG. An in-depth discussion on VSL brominated species and the inorganic bromine budget can be found in Dorf (2005).

Reactions decomposing brominated organics in the stratosphere are very similar to the ones releasing reactive chlorine from its organic precursors. Reactive inorganic bromine species are produced by reactions with OH and $O(^1D)$ and by photolysis, before they can take part in ozone destroying catalytic cycles

$$Br + O_3 \longrightarrow BrO + O_2 \quad\quad (R1.63)$$
$$BrO + O \longrightarrow Br + O_2, \quad\quad (R1.64)$$

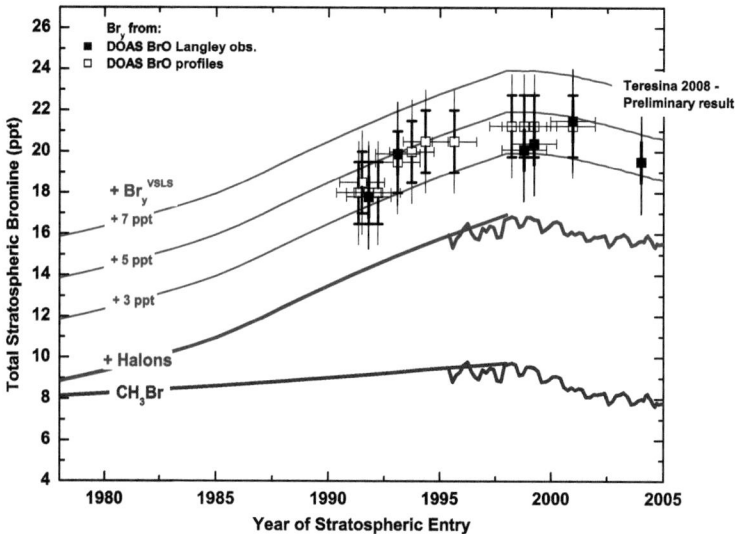

Figure 1.15: Measured trends for bromine (ppt) in the near-surface troposphere (lines) and stratosphere (squares). Global tropospheric bromine from methyl bromide as measured in ambient air and firn air (thin solid line - no correction has been made for tropospheric loss of CH_3Br) Butler et al. [1999] till 1998 and Montzka et al. [2003] past 1995; global tropospheric bromine from the sum of methyl bromide plus halons as measured in ambient air, archived air and firn air (thick solid line) Butler et al. [1999] and Fraser et al. [1999] till 1998 and Montzka et al. [2003] past 1995; and, bromine from CH_3Br and halons plus bromine from VSLS organic bromine compounds near the tropopause or transport of bromine bearing inorganic gases or bromine containing aerosols [Murphy and Thompson, 2000] across the tropopause (BryVSLS) [Salawitch et al., 2005; Pfeilsticker et al., 2000], assuming total contributions of 3, 5, or 7 ppt of these species to Br_y (thin dotted lines). Total inorganic bromine derived from stratospheric measurements of BrO and photochemical modeling that accounts for BrO/Bry partitioning from slopes of Langley BrO observations above balloon float altitude (filled squares) and lowermost stratospheric BrO measurements (open squares). Bold/faint error bars correspond to the precision/accuracy of the estimates, respectively. The years indicated on the abscissa are sampling times for tropospheric data. For stratospheric data, the date corresponds to the time when the air was last in the troposphere, i.e. sampling date minus estimated mean time in stratosphere. Preindustrial levels of CH_3Br were (5.8 ± 0.3) ppt [Saltzman et al., 2004] in the southern hemisphere and 0 ppt for the halons [Reeves et al., 2005]. Adopted from an update of Dorf et al. (2006).

The main impact on ozone destruction by bromine occurs during increased chlorine activation inside the polar vortex, since the combined cycle of BrO and ClO can effectively destroy ozone.

$$BrO + h\nu \longrightarrow Br + O \quad \text{(R1.65)}$$
$$BrO + NO \longrightarrow Br + NO_2 \quad \text{(R1.66)}$$
$$BrO + ClO \longrightarrow Br + OClO \quad 59\% \quad \text{(R1.67)}$$
$$\longrightarrow Br + ClO_2 \quad 34\% \quad \text{(R1.68)}$$
$$\longrightarrow BrCl + O_2 \quad 7\% \,. \quad \text{(R1.69)}$$

The branching ratios are valid for T = 195 K according to Sander et al. (2003). Because of the fast photolysis of BrCl and the collision-induced decay of OClO to Cl and O_2, the reaction of BrO with ClO leads to coupled catalytic bromine-chlorine ozone-depletion. A faster photolysis of the ClO-dimer, indicated by recent studies (Pope et al., 2007), would also affect the efficiency of the ClO-BrO cycle and slow it down. The most important sinks of BrO are its photolysis and reactions with NO, ClO and NO_2. During sunset, BrO concentrations decrease very rapidly, as the reservoir species become more and more abundant. The main reservoir species are formed in reactions with NO_2, HO_2, ClO and CH_2O.

$$BrO + NO_2 + M \longrightarrow BrONO_2 + M \quad \text{(R1.70)}$$
$$BrO + HO_2 \longrightarrow HOBr + O_2 \quad \text{(R1.71)}$$
$$Br + HO_2 \longrightarrow HBr + O_2 \quad \text{(R1.72)}$$
$$Br + CH_2O \longrightarrow HBr + CHO \,. \quad \text{(R1.73)}$$

The subsequent release of BrO_x occurs through photolysis and in the case of HBr through reaction with OH:

$$BrONO_2 + h\nu \longrightarrow Br + NO_3 \quad \text{(R1.74)}$$
$$\longrightarrow BrO + NO_2 \quad \text{(R1.75)}$$
$$HOBr + h\nu \longrightarrow OH + Br \quad \text{(R1.76)}$$
$$HBr + OH \longrightarrow Br + H_2O \,. \quad \text{(R1.77)}$$

Due to the very efficient photolysis and decomposition of the bromine reservoir species, BrO_x is more efficient in destroying ozone than ClO_x. In the absence of light, heterogeneous processes can lead to bromine activation as observed in the presence of polar stratospheric clouds.

Chapter 2

Physics of radiation and molecular absorption

Radiative transfer is the physical phenomenon of energy transfer in the form of electromagnetic radiation (Van de Hulst, 1981). The propagation of this radiation through a medium is affected by absorption, emission and scattering. These processes are essential for the understanding and interpretation of the present work.

Absorption of electromagnetic radiation by molecules is the phenomenon, used to retrieve the integrated number of the particular molecules along the light path (see 4.1). Consequently, the light path has to be known to convert this number into concentrations.

This chapter discusses the basics of radiation and radiative transfer in the Earth's atmosphere as well as the principles of molecular absorption.

2.1 Radiative transfer in the Earth's atmosphere

Max Planck described 1903 the spectral radiance I of a black body at temperature T by the following formula:

$$I(\lambda, T) = \frac{2hc^2}{\lambda^5} \frac{1}{e^{\frac{hc}{\lambda kT}} - 1}. \tag{2.1}$$

This function represents the emitted power per unit area of emitting surface, per unit solid angle, and per unit frequency. Solar radiation reaching the top of the atmosphere can be approximated as black body radiation at 6000 K. According to Planck's law, emission peaks at about 503 nm in the visible wavelength range. The black body spectrum is superimposed by absorption lines (Fraunhofer lines) since radiation emitted from the inner layers of the photosphere is absorbed in the outer ones.

The flow through a very small area element da located at $\mathbf{r}$ in time dt, in the solid angle $d\Omega$ about the direction $\hat{\mathbf{n}}$ in the frequency interval ν to $\nu+d\nu$ in the radiation field can be characterized by the following formula (where polarization is ignored)

$$dE_\nu = I_\nu(\mathbf{r}, \hat{\mathbf{n}}, t) \cos\theta \, d\nu \, da \, d\Omega \, dt \tag{2.2}$$

where θ is the angle that the unit direction vector $\hat{n}$ makes with a normal to the area element. The amount of radiant energy within a cone of solid angle Ω incident on or coming from an element of area dA per unit time and unit wavelength interval is called irradiance and given by

$$E_\lambda = \int_\Omega I_\lambda \cos\theta \, d\Omega. \tag{2.3}$$

The integration of the radiance over the whole sphere (4π) leads to the actinic flux F_λ which then represents the radiant energy flux out of all directions through a spherical surface.

$$F_\lambda = \int_{4\pi} I_\lambda \, d\Omega. \tag{2.4}$$

A quantity, which influences the diurnal and annual variation of many species is the photolysis rate. It is determined by the number of photons available (which is the actinic flux), the ability of a molecule to absorb these photons (the absorption cross section), and the probability that the photon absorption leads to the decomposition of the molecule (quantum yield for photo dissociation). The product of these 3 factors integrated over all wavelengths contributing to the photolysis of the molecule leads to the compounds photolysis rate J (Brasseur and Solomon, 2005).

$$J = \int F(\lambda)\sigma(\lambda, T)\phi(\lambda) d\lambda \tag{2.5}$$

for F given in photons cm^{-2} s^{-1} nm^{-1}, where $\sigma(\lambda)$ is the temperature dependent absorption cross-section and $\phi(\lambda)$ is the quantum yield for the reaction.

2.1.1 Scattering

Light scattering is one of the two major physical processes that contribute to the visible appearance of most objects, the other being absorption which is described in section 2.1.2.
Scattering can be described in the framework of classical electrodynamics. Here it represents an oscillating polarization of a particle caused by the incident electromagnetic wave, which results in a reradiation with certain energy and direction (Van de Hulst, 1981). Depending on the size of the scattering particle, the scattering angle ψ, between incident and scattered light shows some characteristics. Scattering on particles which are small compared to the wavelength of the incident radiation is called Rayleigh scattering. Scattering on larger particles is referred to as Mie scattering. Both Mie and Rayleigh scattering are considered elastic scattering processes, in which the energy (and thus wavelength and frequency) of the light is not changed. In contrast to that, rotational Raman scattering is the inelastic scattering of a photon.
The three different scattering theories, important for the interpretation of our measurements are discussed below.

Rayleigh scattering

Rayleigh scattering displays the limit of Mie's theory for small particles, where the scattered radiation can be approximated as dipole radiation. This effect was first modeled successfully by Lord Rayleigh at

2.1. RADIATIVE TRANSFER IN THE EARTH'S ATMOSPHERE

Figure 2.1: Sunset at the measurement side, in Teresina, Brasil: Rayleigh scattering causes the blue hue of the daytime sky and the reddening of the sun at sunset, the scattering occurring on cloud droplets is explained by Mie theory.

the end of the 19th century. In order for Rayleigh's model to apply, the particle must be much smaller in diameter than the wavelength λ of the scattered wave (<0.1 λ); which is true for air molecules. In this size regime, the exact shape of the scattering center is usually not very significant and can often be treated as a sphere of equivalent volume.
Instead, the molecule is characterized by its polarizability α, which describes how much the electrical charges on the molecule will move in an electric field. In this case, the Rayleigh scattering intensity for a single particle is given by

$$I = I_0 \frac{8\pi^4 \alpha^2}{\lambda^4 R^2}(1 + \cos^2 \theta). \tag{2.6}$$

The amount of Rayleigh scattering of a single particle can also be expressed as a cross section σ_s. Integrating over the sphere surrounding the particle gives the Rayleigh scattering cross section

$$\sigma_s = \frac{2\pi^5}{3} \frac{d^6}{\lambda^4} \left(\frac{n^2 - 1}{n^2 + 2}\right)^2 \tag{2.7}$$

The strong wavelength dependence of Raleigh scattering (λ^{-4}) means that UV radiation is scattered much more readily than visible light. In the atmosphere, this results in blue wavelengths being scattered to a greater extent than longer wavelengths, hence, at daytime one can see blue light coming from all regions of the sky.

Mie scattering

Mie theory is an analytical solution of Maxwell's equations for the scattering of electromagnetic radiation by spherical particles. The Mie solution is named after its developer, German physicist Gustav Mie. In contrast to Rayleigh scattering, the Mie solution to the scattering problem is valid for all possible ratios of diameter to wavelength $\frac{d}{\lambda}$. Mie theory is very important when diameter-to-wavelength ratios are of the order of unity and larger as for scattering on haze and clouds. Mie scattering occurs mostly in the lower layers of the atmosphere where larger particles are more abundant, and dominates when cloud conditions are overcast.

The macroscopic scattering coefficient $k_{\lambda,M}$ of a composite aerosol containing N particles with different radii is defined by

$$k_{\lambda,M} = \int_0^\infty \sigma_{\lambda,M}(\eta)\, n(\hat{r})\, d\hat{r} \qquad (2.8)$$

with $n(\hat{r}) = \frac{dN}{d\hat{r}}$ the particle size distribution with respect to a normalized radius $\hat{r}$ and $\sigma_{\lambda,M}(\eta)$ the scattering cross section. In contrast to Rayleigh scattering, the distribution of scattering angles of Mie scattering (phase function) peaks in forward direction, where for Rayleigh scattering equation (2.6) is found.

The fact, that aerosols also absorb radiation is quantified by the single scattering albedo ϖ_λ,

$$\varpi_\lambda = \frac{k_{\lambda,M}}{k_{\lambda,M} + k_{\lambda,A}}, \qquad (2.9)$$

where $k_{\lambda,A}$ is the absorption coefficient of the particle.

Rotational Raman scattering

When light is scattered from an atom or molecule, most photons are elastically scattered, however, a small fraction of the scattered light is scattered by an excitation. Raman scattering is the inelastic scattering of a photon with the scattered photons having a frequency different from the incident one. In the atmosphere, Raman scattering can occur on nitrogen and oxygen molecules. This is visible in scattered skylight spectra through a filling in of the Fraunhofer lines. The photons measured at the center of the Fraunhofer lines originate from adjacent positions on the edges of the same Fraunhofer line shape.

The rotational Raman scattering cross section has the same wavelength dependence as in the Rayleigh case (i.e. λ^{-4}). The ratio of Raman and Rayleigh scattering lies in the order of some percents. The influence of rotational Raman scattering on scattered light spectra can be quantified quite accurately as discussed in section 4.1.

Surface scattering

A key quantity in the context of atmospheric radiative transfer is the surface albedo ω, describing scattering off the ground. The albedo is defined as the ratio of the reflected (I_r) and the incident intensity (I_i): and is highly variable depending on the surface type. The albedo is also wavelength dependent and typically decreases in the UV (Trishchenko et al., 2003).

2.1.2 Absorption and emission

Absorption of electromagnetic radiation is the way by which the energy of a photon is taken up by matter, in the UV typically by the electrons of an atom, by transition among the quantum mechanical energy levels. Quantum electrodynamics allows to calculate the absorption and emission probabilities by treating the incident electromagnetic wave as a small, time-dependent perturbation to the unperturbed atomic or molecular Hamiltonian (Demtröder, 2000). The transition rates between an upper energy

2.1. RADIATIVE TRANSFER IN THE EARTH'S ATMOSPHERE

level j and a lower energy level i are described in terms of the Einstein coefficients for absorption B_{ij}, stimulated emission B_{ji} and spontaneous emission A_{ji}. The change of the number of molecules in the lower state dn_i depends on the population densities n_i and n_j and the energy density u_λ of the radiation field,

$$\frac{dn_i}{dt} = -n_i u_\lambda B_{ij} + n_j u_\lambda B_{ji} + n_j A_{ji}. \tag{2.10}$$

As a result the energy of the radiation field is changed by $\frac{h}{\lambda}$ with the wavelength λ of the transition and Planck's constant h. The corresponding change in intensity of radiation passing through a small volume of length ds with $n = n_i + n_j$ molecules (per volume) is given by

$$dI_{Abs} + dI_{StE} + dI_{SpE} = -n_i \frac{B_{ij} h c}{\lambda} I_\lambda ds + n_j \frac{B_{ji} h c}{\lambda} I_\lambda ds + n_j \frac{A_{ji} h c}{4\pi\lambda} ds, \tag{2.11}$$

where the factor $\frac{1}{4\pi}$ indicates that spontaneous emission is isotropic while stimulated emission only occurs in propagation direction. Since altitudes below 80 km in the Earth's atmosphere, can be approximated in local thermodynamic equilibrium, the energy levels are populated according to Boltzmann's law and the energy density of the radiation is given by Planck's distribution. Then equation 2.11 reads

$$dI_{Abs} + dI_{StE} + dI_{SpE} =$$

$$= -n_i \frac{B_{ij} h}{\lambda} I_\lambda ds + n_i \frac{B_{ij} h}{\lambda} I_\lambda e^{\frac{-\Delta E}{k_B T}} ds + n_i \frac{B_{ij} h}{\lambda} \frac{8\pi h c}{4\pi \lambda^5} e^{\frac{-\Delta E}{k_B T}} ds =$$

$$= -n_i \sigma_{\lambda,A} \left(1 - e^{\frac{-\Delta E}{k_B T}}\right) \left[I_\lambda - \frac{8\pi h c}{4\pi \lambda^5} \frac{1}{e^{\frac{\Delta E}{k_B T}} - 1}\right] ds = \tag{2.12}$$

$$= -k_{\lambda,A} \left[I_\lambda - \frac{1}{4\pi} P_\lambda\right] ds,$$

with the Boltzmann constant k_B, the temperature T, the energy difference between the two levels ΔE, $\sigma_{\lambda,A} = \frac{B_{ij} h}{\lambda}$ the absorption cross section, $k_{\lambda,A} = n_i \sigma_{\lambda,A} \left(1 - e^{\frac{-\Delta E}{k_B T}}\right)$ the the absorption coefficient and P_λ Planck's distribution of thermal radiation. For temperatures occurring in troposphere and stratosphere the population of the lower energy level can be approximated by the number density of molecules available in the considered volume $n \simeq n_i$. Therefore the radiant energy emitted by stimulated emission is much smaller than the absorbed energy and the exponential term in the definition of the absorption coefficient $k_{\lambda,A}$ can be neglected.

2.1.3 The equation of radiative transfer

The equation of radiative transfer describes the travel of a beam of radiation through the atmosphere by means of its loss of energy due to absorption, its gain of energy due to atmospheric emission and its redistribution of energy due to scattering. Neglecting emission of terrestrial long wave radiation (valid in the UV and visible wavelength range) the differential form of the equation for radiative transfer reduces to Beer-Lambert's law,

$$dI_\lambda = -(k_{\lambda,S} + k_{\lambda,A}) I_\lambda ds = -e_\lambda I_\lambda ds, \tag{2.13}$$

with the extinction coefficient e_λ and combined Rayleigh and Mie scattering coefficients ($k_{\lambda,S} = k_{\lambda,R} + k_{\lambda,M}$). dI_λ represents the change in intensity of a beam of radiation after passing through a small atmospheric volume of length ds. Integrating equation (2.13) leads to

$$I_\lambda(s) = I_\lambda(s_o) e^{-\int_{s_o}^{s} e_\lambda(s')ds'} = I_\lambda(s_o) e^{-\tau_\lambda(s_o,s)} \quad (2.14)$$

This relation between the optical density $\tau_\lambda(s_o, s) = \ln \frac{I_\lambda(s_o)}{I_\lambda(s)} = \int_{s_o}^{s} e_\lambda(s')ds'$ between s_o and s, and the incident radiance $I_\lambda(s_o)$ is the basis for the DOAS method, which is described in chapter 4.1.

2.2 Molecular absorption in the atmosphere

Figure 2.2 shows the actinic flux as modeled with the RTM McArtim in the mesosphere, at 69 km, in the stratosphere, at 29 km altitude and in the troposphere, near the ground at 5 km. The depicted wavelength's range from 200 to 610 nm. The spectrum of the actinic flux at the earth's surface has several components; direct radiation comes straight from the sun, the other contribution is radiation scattered by molecules, aerosols or the surface. All the radiation that reaches the ground passes through the atmosphere, which modifies the spectrum by absorption and scattering, as discussed earlier. Atomic and molecular oxygen and nitrogen absorb very short wave radiation, effectively blocking radiation with wavelengths < 242 nm. When molecular oxygen in the atmosphere absorbs short wave ultraviolet radiation, it breaks due to photolysis. This leads to the production of ozone (see chapter 1.7). Ozone strongly absorbs longer wavelength UV in the Hartley band from 242 to 300 nm and weakly absorbs visible radiation. Wavelength dependent Rayleigh scattering and scattering from aerosols and other particulates, including water droplets, also change the spectrum of the radiation that reaches the ground (and make the sky blue).

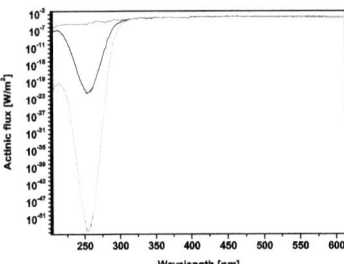
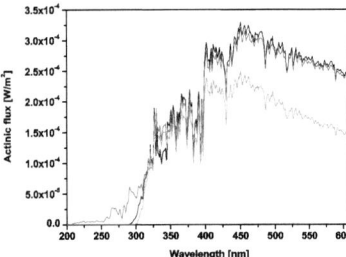

Figure 2.2: Actinic fluxes simulated with McArtim (Deutschmann, 2008) for 5 km (green), 29 km (black) and 69 km (red) height on a clear day. Left panel: On a logarithmic scale, highlighting the absorption between 200 and 300 nm, which is mainly due to ozone and oxygen. Right panel: On a linear scale, highlighting the visible wavelength range, where the decrease at 5 km is mainly due to Mie scattering.

2.2. MOLECULAR ABSORPTION IN THE ATMOSPHERE

Figure 2.3: Simulated radiative transfer events of balloon-borne limb scattered skylight measurements with our mini-DOAS instrument in 34 km altitude and elevation of the telescope $\alpha = -2°$. The coloured dots denote photon - matter interaction, Rayleigh scattering (red), Mie scattering (green), absorption (blue) and ground scattering (yellow).

Figure 2.3 illustrates the influence of scattering events on the measurement geometry of the here presented measurements. For the simulation the detector is placed at 34 km altitude and the elevation of the telescope is $\alpha = -2°$. The coloured dots denote photon - matter interaction, where Rayleigh scattering is shown in red, Mie scattering in green, absorption in blue and ground scattering is displayed yellow. The detector is located at the red points point of divergence. Since the RTM is modelled backward, from the detector to the sun, absorption events characterize the border of a trajectory, while all other events can be reversed.

Chapter 3
Instrumental design and performance

The mini-DOAS instrument used for this study was designed (during an earlier Phd-thesis) at the IUP Heidelberg and is discussed in Weidner (2005). As there were only small modifications, a brief description is given here.
The mini-DOAS spectrometer was designed for low weight ($\approx$ 7kg) and low power consumption (7.5 W), with particular emphasis on stable optical imaging and a reasonably large signal to noise ratio. While the former characteristic offers the chance for versatile applications, the latter feature is found to be necessary for the detection of O_3, NO_2 and in particular of the weakly absorbing gases (e.g. OClO, BrO, HONO, OIO, or IO), as based on the experience made in the stratosphere with the direct sunlight DOAS (e.g.Ferlemann et al. (1998); Harder et al. (1998)).

3.1 Design

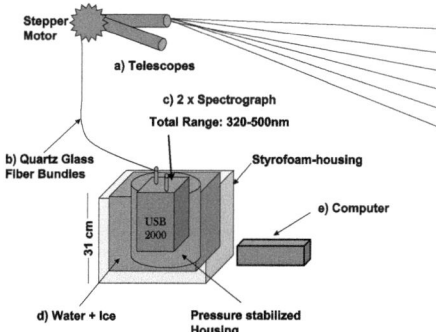

Figure 3.1: Sketch of the mini-DOAS instrument components. A detailed description is given in the text.

The mini-DOAS instrument consists of five major parts (see Figure 3.1): (a) Two light intake telescopes, ending in (b) glass fiber bundles which conduct the skylight from the telescopes into the spectrometers, (c) two commercial Ocean Optics USB-2000 spectrometers, housed in a (d) sealed aluminium housing for pressure stabilization and put into a water-ice-bath for temperature stabilization, and finally (e) a single board computer for data handling and storage.

(a) Each of the two telescopes consists of a spherical quartz lens (12.7 mm in diameter, 30 mm focal length), which focuses the incoming scattered skylight onto the round or rectangular entrance of a glass fiber bundle. During a balloon flight, the telescopes are mounted to the structure of the particular gondola, where one telescope is mounted on an elevation angle scanner (built by Hofmann Meßtechnik, Rauenberg, Germany), which supports Limb observations between +10 ° and -20 ° elevation, with step sizes as small as 0.04 °.

(b) Each glass fibre bundle consists of 11 individual quartz glass fibers (100 μm in diameter, 2 m in length, numerical aperture NA = 0.22). They allow for a flexible arrangement of the instrument inside the particular balloon gondola, but also reduce the polarization sensitivity of grating spectrometers (Stutz and Platt, 1997). This helps to keep the intensity stable for changing azimuth directions. In this case the polarization sensitivity of the Ocean Optics USB 2000 spectrometer, by using glass fibre bundles, is ≈ 1 % (Weidner, 2005). As one of the two channels was initially used for nadir observations, the individual glass fibres are arranged in round geometry at the light intake, a mounting which supports a field of view (FOV) of 0.8 ° when combined with the telescope. For the second channel, which was from the very beginning determined for Limb observations, the glass fibres are arranged in a rectangular geometry light intake setup i.e., the individual glass fibre entrances are aligned linearly. This arrangement supports a FOV of 0.19 ° in the vertical and 2.1 ° in the horizontal direction. Likewise, the glass fibres are linearly aligned at both exits, and illuminate the 50 μm wide and 1000 μm high spectrometer entrance slit completely. Figure 3.2 shows a sketch assessing the light throughput and FOV of the combination of fiber and telescope.

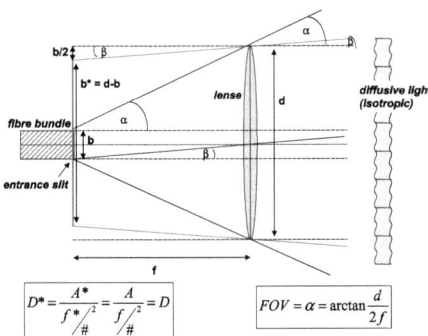

Figure 3.2: Sketch of the optics governed by the telescope and the fibers. Note, that the light throughput D is independent of the lens for isotropic radiation, while the FOV is altered by the use of a lens. [Sketch from Bodo Werner, personal communication].

(c) The main part of the mini-DOAS balloon instrument consists of two commercial Ocean Optics USB

2000 spectrometers. The USB 2000 is a miniature grating spectrometer working in crossed Czerny-Turner geometry. It is of small size (86 x 63 x 30 mm ^{3}), low weight (270 g) and provides a high photon detection sensitivity due to an integrated linear CCD array detector (Sony ILX511). The light enters the spectrometer through an entrance slit (50 µm × 1000 µm) from where it is focused by a collimator mirror onto a holographic grating with 1800 grooves/mm. A second mirror focuses the light onto the linear CCD array with 2048 pixels (each pixel is 14 µm wide and 200 µm high). Attached onto the CCD array detector is a cylinder lens, which focuses the 1000 µm high entrance slit onto the 200 µm high detector. Also attached to the CCD array detector is the preamplifier and a control logic unit, which handle the pre-amplification of the analogous signals, A/D conversion to 12 bit data, and communication.
The spectrographs cover a spectral range of approx. 330 - 550 nm at a full width at half maximum (FWHM) resolution of 0.8 - 1.0 nm, or 8 - 10 pixel/FWHM, depending on the wavelength. The wavelength coverage and resolution allow for the detection of the atmospheric trace gases O_3, NO_2, O_4, H_2O, BrO, and OClO (and potentially IO, OIO, CH_2O).
(d) Both spectrometers are kept in a pressure stabilized housing made of aluminum, which enables the instrument to maintain a pressure below 10^{-1} hPa for more than a week. For temperature stabilization the whole aluminum pot with the spectrographs inside sits in an epoxy glass housing filled with a water ice mixture. This ensures a stable spectrometer and CCD array temperature of $0°C$ during the entire balloon flight.
(e) Data handling and storage is maintained by a single board PC (Lippert: "cool road runner") equipped with a flash memory. The allocated data are transferred from the spectrometers to the PC via a USB data transfer connection. The PC is operated under Windows XP with our lab-owned DOASIS (Kraus, 2004a) or, recently, MSDOAS (Udo Friess, private communication) softwares packages. Both software tools support the automatic adjustment of the integration time, and the recording and storage of the measured spectra.
The total size of the instrument is 260 x 260 x 310 mm ^{3} (w/o fibers), its weight is around 4.3 kg (including fibers, stepper motor and electronics) plus 2.8 kg of water-ice-mixture, and its power consumption is around 7.5 W.

3.1.1 Wavelength dependency of the FOV

The wavelength dependency of the refraction index of the applied lens leads to a certain dependency of the FOV on the wavelength. The relevance of this effect to our measurements is estimated in the following. As illustrated in Figure 3.2, the FOV α is given by

$$\alpha = arctan\frac{D}{2f} \quad (3.1)$$

The relation of the focal length to the refraction index n is

$$f = \frac{1}{(n-1)} \cdot \frac{1}{(\frac{1}{r_1} + \frac{1}{r_2})} \quad (3.2)$$

where r_1 and r_2 are the radii of curvature of the lens. The error of f with respect to n reads

$$\Delta f = \frac{df}{dn} \cdot \Delta n = \frac{-1}{(n-1)^2} \cdot \frac{1}{(\frac{1}{r_1} + \frac{1}{r_2})} \cdot \Delta n \quad (3.3)$$

The refraction index of suprasil, which is the material of the lenses used for this study, is given by the manufacturer for different colors/wavelength:
n = 1.45846 at 587.6 nm
n = 1.46313 at 486.1 nm
n = 1.46669 at 435.8 nm
Assuming a linear dependency of the refraction index on the wavelength, a change of wavelength can be associated with a change in the FOV. The wavelength range of the used spectrograph is around 200 nm, which corresponds to a Δn of 0.01 nm. The resulting Δf is 0.5.
The error of the FOV with respect to f reads

$$\Delta FOV = \frac{dFOV}{df} \cdot \Delta f = \frac{-1}{1+(\frac{d}{2f})^2} \cdot \frac{d}{(2f)^2} \cdot \Delta f \tag{3.4}$$

The resulting error for the FOV is $0.006°$. Compared to the FOV in the vertical of $0.19°$ it is negligible.

3.2 Performance

3.2.1 Instrumental noise

As already shown by Weidner (2005), the most dominant source of instrumental noise is the statistical noise of the photon-electrons, which limits the residual physically to 0.67 % for 1 scan with 80 % saturation. The second most important noise contribution is caused by the detector and its attached electronics. They contribute to the residual with 0.095 %. For typical integration times of around 100 ms during balloon flights, the dark current noise only contributes with 0.004 % to the total noise, and is therefore regarded negligible. The root-mean-square sum of the above contributions gives a total noise of 0.68 %. Since all these sources of noise are purely statistical and proportional to the number of scans, the total noise of the measurement can be reduced by summing up multiple spectra. During balloon flights up to 1000 spectra (corresponding to around 20 s integration time) are co-added, resulting in a theoretical total noise of 0.024 %. Figure 3.3 shows the total noise as a function of the number of co-added spectra for the Limb spectrograph. The number of counts as shown in the left panel is a result of subtracting one spectrum from the another, both recorded at identical conditions, the residual shown in the right panel is a result of performing an DOAS analysis (see section 4.1).

The most important systematic source of noise is the diode structure, caused by the different sensitivities of the individual detector pixels. For a completely stable instrument this would cause no problems, as every spectrum is divided by a reference spectrum during the DOAS evaluation process. However, as soon as there is a spectral shift due to pressure and temperature instability inside the instrument, every single pixel is not divided by the same detector pixel of the reference spectrum. The resulting structures are eliminated by dividing every spectrum by a high pass filtered white lamp spectrum prior to the DOAS evaluation.

3.2. PERFORMANCE

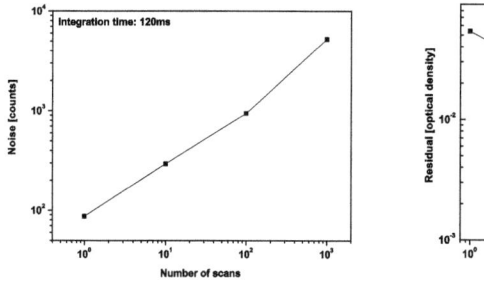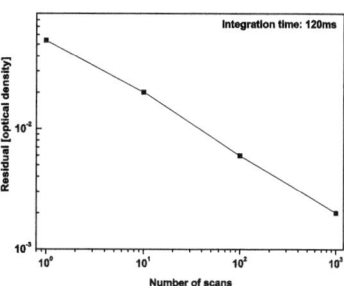

Figure 3.3: Left panel: Total noise as a function of co-added spectra at the 80% illumination level of the CCD array detector for skylight. Right panel: Peak to peak residual of the same spectra.

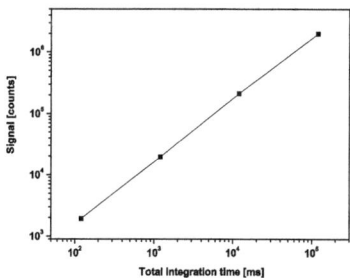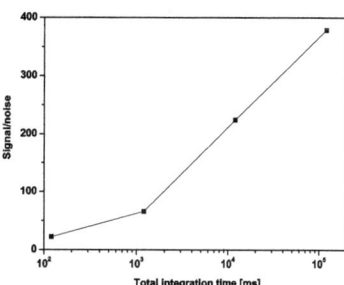

Figure 3.4: Left panel: Signal as a function of total integration time at constant illumination. Right panel: The ratio of signal to total noise as a function of total integration time.

3.2.2 Observation geometry

Definition
The viewing geometry is defined by the position and orientation of the telescope and the position of the sun, both relative to the earth coordinate system (where the atmosphere is defined). The position of the sun can be calculated for a given time.

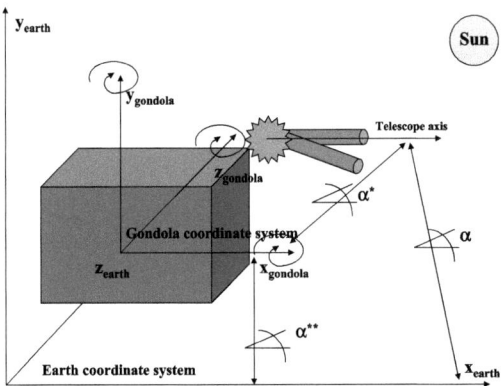

Figure 3.5: Illustration of the mini-DOAS attitude with respect to the earth coordinate system, by introducing an additional coordinate system in which the gondola is fixed (gondola coordinate system), where moving of the telescope is only governed by the mini-DOAS instruments controll system. The resulting angles α^{**} (between $x_{gondola}$ and x_{earth}), α^{*} (between $x_{gondola}$ and the telescope optical axis) and α (between x_{earth} and the telescope optical axis), which is the sum of both and an important input for RTM.

The telescope is fixed to the particular balloon gondola as displayed in figure 3.5. The angle α^{*} (between $x_{gondola}$ and the telescope optical axis) is controlled by commands of the measurement routine, saved on the computer, but holds a possible offset $\Delta\alpha^{*}$. A description on how α^{*} is measured is given in section 3.2.3 and on how it is retrieved in section 4.3.

The position of the gondola (latitude and longitude) and the angle between z_{earth} and $z_{gondola}$ (the azimuth angle of the gondola relative to north) is given by the particular gondola attitude system.

The possible pendulum oscillation of the gondola around $z_{gondola}$ is an output of the attitude system which is only given by certain balloon gondolas like for the MIPAS-B payload . The related angle is referred to as α^{**}. A characterisation on how gondola oscillations around α^{**} influence the measurements is given in section 4.3.

The sum of α^{*} and α^{**} is called the elevation angle α (between x_{earth} and the telescope optical axis). The balloon-borne Limb scattered skylight measurements are most sensitive to the elevation angle α, which is the parameter allowing for the height resolution of the measurements (see section 4.2). If it is not known, it is responsible for the largest error in the resulting concentration profiles (see section 4.3) Routinely the telescope is oriented to the horizon at fixed elevation angle α^{*} during balloon ascent as

3.2. PERFORMANCE

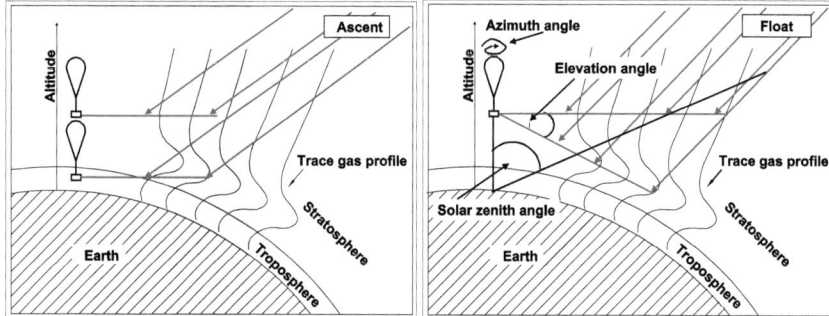

Figure 3.6: Viewing geometry of balloon borne Limb scattered skylight measurements during balloon ascent (left panel) and during float (right panel).

shown in figure 3.6 (left panel). When the gondola dives through the atmosphere the instrument provides a large sensitivity for trace gases in the particular tangent altitude.
A second mode of operation is started when the balloon reaches float altitude (e.g. around 35 km, see figure 3.6, right panel). Then the telescope is commanded to automatically scan different elevations ranging from $\alpha = 0°$ to $\alpha = -6°$ in steps of $0.5°$. The change in observation geometry leads to a changing sensitivity in different altitudes of the atmosphere, with the highest sensitivity at flight altitude (for $\alpha \approx 0°$) or 10 km below (for $\alpha \approx -5°$). In practice, by scanning through the atmosphere, the instrument provides time-resolved profile information on UV/vis absorbing trace gases. Thus Limb scattered skylight observations of the mini-DOAS are more flexible in observation geometry, compared to direct sun measurements as illustrated in figure 3.6.
In order to get altitude resolved information of the current atmosphere with direct sun measurements the relative position of the balloon and the sun compared to the atmosphere has to change. This is only the case during balloon ascent, and during sunrise or sunset. For Limb scattered skylight measurements profile information is available during ascent, when the elevation angle is fixed in near horizontal position and during float, when the balloon has reached a constant altitude, the atmosphere can be probed by Limb scanning, as described above.
In order to provide a constant local SZA along most of the light path, the telescope is fixed to an azimuth angle of 90° relative to the sun (given that the relative azimuth of the gondola is constant, which can not always be guaranteed). This is important when measuring photochemically active species like BrO or NO_2 mainly during sunrise or sunset when their concentrations change rapidly.

3.2.3 Flight preparations

As described above, the relative position of the telescope to the gondola (α^*) affects the absolute elevation angle α. Therefore the telescope is carefully aligned to the principle axis of the payload prior to the balloon flight. This is done on a hanging and balanced gondola, so that the relative position of the gondola to the system earth-atmosphere resembles flight conditions. The horizontal position of the telescope is defined by triangulation with a laser pointer fixed to the telescope as illustrated in Figure 3.7.

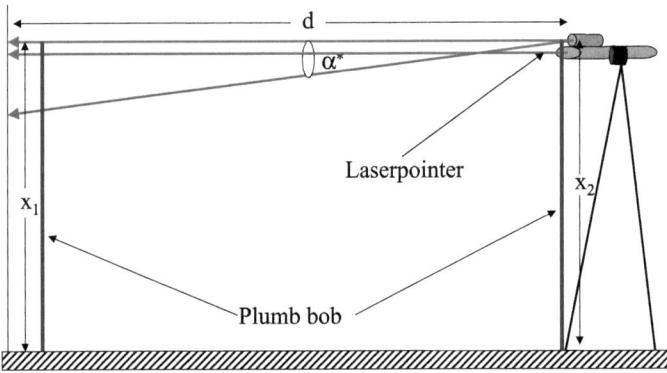

Figure 3.7: Sketch of the triangulation with the laser pointer to fix the elevation angle of the telescope prior to the flight, d, x_1, x_2, α^* as explained in the text.

The height of the telescope above ground is measured using a plumb bob and brought into agreement with the height of the laser point on a wall by subsequently moving the motor of the telescope. With $\alpha^* = 0$, for $x_1 = x_2$. The triangulation error is the given by

$$\Delta x = x_1 - x_2 \qquad (3.5)$$
$$\Delta \alpha = arctan(\Delta x / d) \qquad (3.6)$$

where d is the distance to the wall and Δx the error of the height measurement, which can be caused by an uneven ground and bad alignment. Assuming a distance of d = 10 m and a Δx = 5cm results in an elevation error of $\Delta \alpha^* = 0.3°$. Another source of error in the course of this procedure is a possible misalignment of the laser pointer to the telescope. This is assessed to another $\Delta \alpha^* = 0.3°$ resulting in a total pointing error of $\Delta \alpha^* = 0.6°$.

A remaining misalignment can be tested after the flight by comparing the modeled and measured relative radiances for each observation. This procedure serves as a sensitivity test for pointing, since the radiance of skylight in the UV/vis spectral range near the horizon largely changes with tangent height and shows a wavelength depend maximum in the lowermost stratosphere (Sioris et al., 2004; Weidner, 2005). The method is described in chapter 4.

Among a larger series of stratospheric balloon flights during which we deployed the mini-DOAS instrument since 2002 (see Table: 3.2.3), measurements presented in this work are from deployments of the mini-DOAS at Teresina on June 14, 17, and 30 in 2005 (see section 5).

3.2. PERFORMANCE

Table 3.1: Compendium of balloon-borne mini-DOAS measurements. The last eight flights were conducted within the present thesis.

Date Time (UT)	Location	Geophys. Cond. SZA range	Instrument payload	Observation Mode
18/19 Aug. 2002 15:15–2:38	Kiruna 67.9° N, 21.1° E	high lat. sum. 69.75–94.4° 94.6–88.1°	LPMA/ mini-DOAS	nadir fixed Limb
4 March 2003 12:55–15:25	Kiruna 67.9° N, 21.1° E	high lat. spring 77.6–88.8°	LPMA/DOAS mini-DOAS	nadir fixed Limb
23 March 2003 14:47–17:35	Kiruna 67.9° N, 21.1° E	high lat. spring 78.9–94.7°	LPMA/DOAS mini-DOAS	nadir fixed Limb during ascent Limb scanning at float
9 Oct. 2003 15:39–17:09	Aire-sur-l'Adour 43.7° N, 0.25° W	mid-lat fall 66–88°	LPMA/DOAS mini-DOAS	nadir fixed Limb
24 March 2004 13:55–17:35	Kiruna 67.9° N, 21.1° E	high lat. spring 72–98°	LPMA/DOAS mini-DOAS	fixed Limb during ascent Limb scanning at float
14 June 2005 9:00–17:10	Teresina 5.1° S, 42.8° W	tropical winter 93–29°	mini-DOAS MIPAS-B	Limb scanning at float
17 June 2005 9:00–17:10	Teresina 5.1° S, 42.8° W	tropical winter 61–95°	mini-DOAS LPMA/DOAS	fixed Limb during ascent Limb scanning at float
30 June 2005 9:00–17:00	Teresina 5.1° S, 42.8° W	tropical winter 93–29°	mini-DOAS LPMA/IASI	Limb scanning at float
1 March 2006 8:00–17:22	Kiruna 67.9° N, 21.1° E	high lat. spring 77–100°	LPMA/IASI mini-DOAS	fixed Limb during ascent Limb scanning at float
13 June 2008 8:45–13:50	Teresina 5.1° S, 42.8° W	tropical winter 93–29°	mini-DOAS LPMA/IASI	Limb scanning at float
27 June 2006 9:00–21:18	Teresina 5.1° S, 42.8° W	tropical winter 60–95°	mini-DOAS LPMA/DOAS	fixed Limb during ascent Limb scanning at float
5 March 2009 3:34–5:10	Kiruna 67.9° N, 21.1° E	high lat. spring 77–100°	LPMA/IASI mini-DOAS	Limb scanning at float
7/8 September 2009 14:50–6:04	Kiruna 67.9° N, 21.1° E	high lat. spring 77–100°	LPMA/DOAS mini-DOAS	fixed Limb during ascent Limb scanning at float

Chapter 4

Retrieval methods

As there are many steps from the primarily measured quantities (spectral intensities) to the final product (time series of trace gas profiles), care is needed to quantify the shortcomings at each step and the propagation of those into the next. This chapter provides a detailed description of the applied retrieval methods and a characterization of their information content and error budget. Figure 4.1 displays a

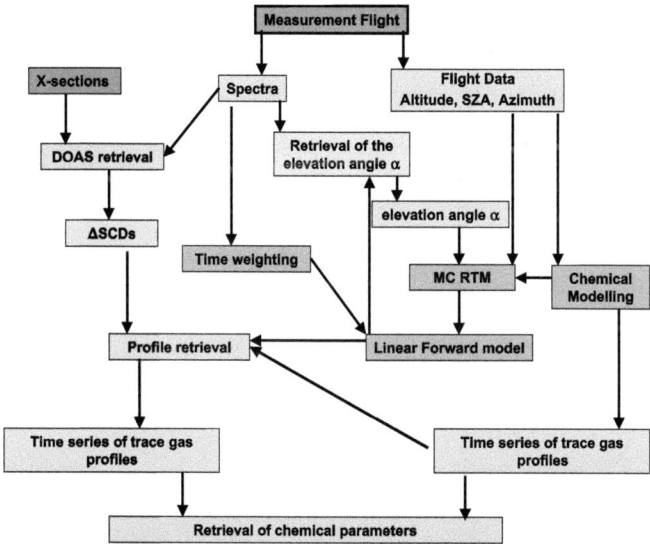

Figure 4.1: Schematic drawing of the procedure applied to retrieve time series of trace gas profiles and chemical information from mini-DOAS measurements. Input and output parameter are shown in yellow, modeling is colored blue and retrievals are shown in green (for some boxes this categorization is ambiguous).

schematic drawing of the procedure applied to retrieve time series of trace gas profiles and chemical information from mini-DOAS measurements. The spectral retrieval is performed by the fitting routine of WinDOAS (Fayt and van Roozendael, 2001) for chemical modeling the group internal 1-D model Labmos is used, while all other steps are conducted by self written matlab routines.

4.1 Spectroscopic analysis

The spectroscopic analysis of the measured intensity spectra is performed by applying the Differential Optical Absorption Spectroscopy (DOAS) (Platt et al., 1979; Platt and Stutz, 2005), which was mainly developed around 1979 at the FZ Jülich and has evolved into a versatile method for atmospheric remote sensing. Earlier spectroscopic approaches were used to measure O_3 and NO_2, e.g by Noxon (1975).

The general approach for the spectral retrieval is to define a state vector $\mathbf{x}$, containing the parameters to be retrieved, and a forward model $\mathbf{F}$, containing the physics to model the measurement vector $\mathbf{y}$. The equation to solve is given by

$$\mathbf{y} = \mathbf{F}(\mathbf{x}, \mathbf{b}) + \epsilon, \tag{4.1}$$

where ϵ is the measurement error and $\mathbf{b}$ is a vector, that comprises all forward model parameters, influencing the forward model, but will not be retrieved. In the case of DOAS the measurement vector $\mathbf{y}$ consists of measured intensities at different wavelength, which is the observed spectrum $I(\lambda)$ in a chosen wavelength range. The state is represented by the amount of absorption of a certain trace gas along the line-of-sight and forward model parameters are e. g. the absorption cross sections. Given the measurements $\mathbf{y}$ and the forward model parameters $\mathbf{b}$, an inverse problem has to be solved in order to retrieve the state vector $\mathbf{x}$ (Rodgers, 2000).

In order to set up a forward model, the connection between measurement y and state x is explained in the following. The information about the trace gases in the atmosphere is stored in the measured spectra by the absorption structure in the incoming radiation along the line-of-sight. Identification of a certain trace gas is possible because of the specific absorption structure and quantization is based on its strength. What we actually learn from the spectral retrieval is the Slant Column Density (SCD), which is the integrated number of molecules of a certain absorber along the light path of the measure photons.

Several approximations allow for the construction of a forward model $\mathbf{F}$ based on quantities measured by our spectrograph.

Extinction of electromagnetic radiation by matter is described by Lambert-Beer's law (see equation 2.14):

$$I(\lambda) = I_0(\lambda) e^{-\sigma(\lambda) \cdot n \cdot L} \tag{4.2}$$

with $\sigma(\lambda)$ the molecular absorption cross sections, n the number of molecules along the line-of-sight L and $I_0(\lambda)$ the "top-of-the-atmosphere" solar spectrum. Here, for simplicity only one type of absorbing molecule is considered.

The goal of the spectral retrievals is to retrieve the number density of absorbing molecules integrated along the line-of-sight which is referred to as Slant Column Density (SCD),

$$SCD = \int_L n \, ds. \tag{4.3}$$

Replacing $n \cdot L$ in Equation 4.2 leads to

$$I(\lambda) = I_0(\lambda) \cdot e^{-\sigma \cdot SCD} \tag{4.4}$$

4.1. SPECTROSCOPIC ANALYSIS

In principle the exact knowledge of all extinction processes along the line-of-sight is necessary to retrieve the SCD of a certain absorber. Alternatively many effects like the strong Fraunhofer lines, caused by absorption in the suns atmosphere are incorporated by analysing the ratio of two different spectra. Those two different spectra are considered as reference spectrum I_{ref}, with little absorption and measurement spectrum I_i:

$$I_{ref}(\lambda) = I_0(\lambda) \cdot e^{-\sigma(\lambda) \cdot SCD_{ref}} \qquad (4.5)$$

Expanding equation 4.5 with I_{ref} from equation 4.4 leads to

$$\begin{aligned} I_i(\lambda) &= \frac{I_0(\lambda)}{I_{ref}(\lambda)} I_{ref} \cdot e^{-\sigma(\lambda) \cdot SCD_i} \\ &= I_{ref}(\lambda) \cdot e^{-\sigma(\lambda) \cdot SCD_{ref}} \cdot e^{\sigma(\lambda) \cdot SCD_i} \\ &= I_{ref}(\lambda) \cdot e^{\sigma(\lambda) \cdot [SCD_i - SCD_{ref}]} & (4.6) \\ &= I_{ref}(\lambda) \cdot e^{\sigma(\lambda) \cdot \Delta SCD_i} & (4.7) \end{aligned}$$

The quantities retrieved from this ratio are the differences of each SCD of the particular gas, henceforth called $\Delta SCDs$, which actually built the state vector **x**.

The fundamental DOAS principle is the separation of the broad band extinction (by scattering due to aerosols and air molecules) and differential structures, caused by the absorption from trace gases.

$$\tau(\lambda) = \int_L \sigma(\lambda) n + k_S(\lambda) \, ds, \qquad (4.8)$$

where n represents the number density of absorbers and the integration is performed along the line-of-sight L.

The spectra are recorded by a charge coupled device (CCD) detector which in general shows a temperature dependent dark signal caused by the thermal discharge of the photo diodes. In addition, prior to digitalization an offset signal is added to the spectra to avoid negative values. Assuming the temperature of the CCD to be constant, dark and offset signal are constant with time. They are measured separately (e.g. when the instrument is not exposed to any light) and subtracted from the measured spectra (Weidner, 2005). The following arguments will be based on dark and offset signal corrected spectra (see section 3.2.1)

Also the molecular absorption cross sections are split into a low and a high frequency component, σ_b and σ_d respectively:

$$\sigma = \sigma_b + \sigma_d \qquad (4.9)$$

where the indices 'b' and 'd' symbolize 'broadband' and 'differential'. Accordingly, equation (4.8) reads

$$\tau(\lambda) = \int_L \sigma_b(\lambda) n + k_S(\lambda) \, ds + \int_L \sigma_d(\lambda) n \, ds = \tau_b(\lambda) + \tau_d(\lambda) \qquad (4.10)$$

with $\tau_b(\lambda)$ the broadband optical density and $\tau_d(\lambda)$ the differential optical density. When assuming the molecular absorption cross sections σ_d to independent of temperature and pressure, then they are also independent of the line-of-sight and the differential optical density is given by

$$\tau_d(\lambda) \simeq \sigma_d(\lambda) \int_L n \, ds = \sigma_d(\lambda) \, \Delta SCD. \qquad (4.11)$$

This approximation allows to infer the ΔSCD of a measurement spectrum with respect to a reference spectrum without knowing the line-of-sight of both. Considering the definitions above, Beer-Lambert's law can be written as

$$I(\lambda) = I_{ref}(\lambda) e^{-(\tau_b(\lambda) + \tau_d(\lambda))} = I_{ref,b}(\lambda) e^{-\tau_d(\lambda)}, \quad (4.12)$$

where $I_{ref,b}(\lambda)$ comprises the reference spectrum and all broadband extinction processes, like Rayleigh and Mie scattering.

As the spectrograph modifies the incoming solar spectrum corresponding to the instrumental characteristics, the measured spectrum $I^*(\lambda)$ is given as a convolution of the incoming signal $I(\lambda)$ with the normalized instrumental line shape function $g(\lambda)$,

$$I^*(\lambda) = \int_{-\infty}^{\infty} I(\lambda') g(\lambda - \lambda') d\lambda' = I(\lambda) \otimes g(\lambda). \quad (4.13)$$

Due to the finite spectral resolution of the spectrograph, the logarithm of the measured spectrum is given by

$$\ln I^*(\lambda) = \ln\left[\left(I_{ref,b}(\lambda) e^{-\tau_d(\lambda)}\right) \otimes g(\lambda)\right]. \quad (4.14)$$

Provided that the differential optical density is small ($\tau_d(\lambda) << 1$) and the variation of $I_{ref,b}$ with wavelength is negligible ($I_{ref,b}(\lambda) = const$), equation (4.14) can be linearized:

$$\ln I^*(\lambda) \simeq \ln\left[I_{ref,b}(\lambda) \otimes g(\lambda)\right] - \tau_d(\lambda) \otimes g(\lambda) \quad (4.15)$$

and constitutes the base of the forward model **F**. Nevertheless the assumptions, that are necessary for the linearisation of the forward model, have to be considered as possible sources of systematic errors. Wavelength ranges for the spectral retrieval are chosen so that the differential optical densities are small (10^{-3} to 10^{-1}) and the approximation $\tau_d(\lambda) << 1$ is justified. The assumption of σ being constant with p is mostly fulfilled for atmospheric pressures, but the molecular absorption cross sections are is generally not independent of temperature, which is a serious shortcoming of the DOAS approach. However, it can be overcome by the use of several (mostly two mathematically orthogonal) cross sections of the same species obtained at different temperatures in the spectral analysis. The dependency of the reference spectrum I_{ref} and accordingly $I_{ref,b}$ of the wavelength leads to an error of Huppert (2000). Section 4.1.2 discusses these systematic errors and possible corrections in detail.

4.1.1 The DOAS forward model

The forward model **F** used for the approximation of ln(I) with several absorbers m in the UV/visible is then given by

$$F(\lambda) = \ln\left[I_{ref}(\lambda, d_{0,ref}, d_{1,ref}, \ldots)\right] + P(\lambda, p_0, \ldots, p_m) - \sum_{m=1}^{N} a_m \sigma_{m,d}(\lambda, d_{0,m}, d_{1,m}, \ldots), \quad (4.16)$$

where $I_{ref}(\lambda, d_{0,ref}, d_{1,ref}, \ldots)$ is the reference spectrum, in the case of the present study, measured by the mini-DOAS instrument during balloon float at high Sun when absorption by atmospheric constituents is minimal. The use of a solar spectrum actually measured by the spectrograph guarantees the correct

4.1. SPECTROSCOPIC ANALYSIS

representation of the solar Fraunhofer lines in the forward model. The absorption cross sections $\sigma_{m,d}$ and $I_{ref}(\lambda, d_{0,ref}, d_{1,ref}, \ldots)$ are forward the model parameters **b**, while the N scaling factors a_m, the M polynomial coefficients P and the additional parameters $d_{j,i}$, (j = 0...m) built the state vector **x**. The resulting amplitude a_m then correspond to the $\Delta \mathrm{SCD}_i$ of the respective absorber. The broad band structures are approximated by a polynomial $P(\lambda, p_0, \ldots, p_M)$ of degree M, which usually ranges between 2 and 5. The additional parameters $d_{j,i}$ quantify a shift and squeeze for the reference spectra I_{ref} and $\sigma_{m,d}$ in order to compensate possible differences in the wavelength-pixel mapping of I_{ref} and $\sigma_{m,d}$ compared to I_i. A possible misalignment of the different spectra is a result of different measurement conditions (e.g., ambient temperature, pressure) which can hardly be avoided. Especially when using a highly structured light source like the sun, the shift and stretch are crucial parameters of $I^*_{ref}(n)$ with respect to $I^*(n)$. Here I(n) accounts for the simplification of the incoming continuous signal by the detector, mapping it on a discrete grid with a finite number of pixels:

$$I^*(\lambda) \rightarrow I^*(n). \tag{4.17}$$

In the spectral analysis, a linear least square fit is applied to derive the parameters a_m and P and a non-linear Levenberg-Marquardt fit to determine the shift and squeeze parameters. A cost function is set up

$$\chi^2 = \sum_{n=1}^{W} \left(\frac{\ln I^*(n) - F(n)}{\epsilon_n} \right)^2 \tag{4.18}$$

that has to be minimized. Here, W is the number of pixels of the spectral range used for the retrieval and ϵ_n is the measurement error of the nth diode resulting from the measurement noise. Usually, a constant measurement error is assumed for all diodes, i.e. $\epsilon_n = \epsilon =$ const. The fitting procedure starts with a linear least-square fit with initial values for $d_{j,i}$. The retrieved values for a_m and P are then input parameters for an iterative Levenberg-Marquardt fit. Only one iterative step is performed and new values for $d_{j,i}$ are obtained, which in turn are used for a new call of the linear fit. The result of the linear fit is again used for a new call of the non-linear fit until the result converges or a certain number of iterations is fulfilled. To account for instrumental stray light caused by reflections inside the spectrograph by light of the 0th or 2nd and higher orders of the grating, an intensity offset O(n), which is a polynomial of up to 2nd order, is introduced. The product of O(n) and the mean intensity I of the spectrum is directly subtracted from the measured intensity, before taking the logarithm. Therefore O(n) is also called prelogarithmic offset. Its coefficients are additional parameters of the non-linear fit.

4.1.2 Characterization of the spectral retrieval and error analysis

According to equation (4.1), errors of the state **x** are a consequence of the retrieval noise due to measurement errors, errors of the forward model parameters, errors due to correlations of the retrieval parameters and errors of the forward model (Rodgers, 2000). Those errors and possible corrections are discussed in the following.

<ins>Retrieval noise due to measurement error</ins>
As discussed in Section 3.2.1, the measurement error of the mini-DOAS instrument is governed by the photo-electron shot noise ϵ_{ph} of the detector. As a result the statistical errors of the fitted parameters and the theoretical detection limit are determined by the noise of the measurement. The statistics of the

number of electrons generated by the photons illuminating a detector pixel is described by a Poisson distribution. Accordingly, the photo-electron shot noise ϵ_{ph} is given by $\sqrt{N_{ph}}$, with N_{ph} the number of photo-electrons. The fitting routine provides the standard deviations of the retrieval parameters as output, in a sense that the true value of parameter x_l lies with 68.3% probability within the boundaries of the given error. The retrieval noise is appropriately represented by the fitting errors only if the residual spectrum $R(n) = \ln I^*(n) - F(n)$ consists of pure noise. As the residual structures in the presented retrievals show groups of correlated neighboring pixels, the fitting errors have to be corrected by a factor described in Stutz and Platt (1996).

Forward model parameter error
Forward model parameters b are the molecular absorption cross sections and the reference spectrum. The principal errors of the molecular absorption cross sections originate from uncertainties of their absolute magnitude and their temperature dependence, which are commonly given by their authors. The molecular absorption cross sections used for this work are listed in Table 4.1.3 including their corresponding errors. The errors of the molecular absorption cross sections are added to the fitting errors through Gaussian error propagation.

Correlations of retrieval parameters
A source of large systematical errors are correlations between the absorption cross sections included in the fit. However, they are hard to describe quantitatively. In fact, the residual structure as a measure of the quality of the DOAS evaluation can even become smaller when structures are improperly fitted by correlating cross sections. For example, correlations can occur between weakly structured absorbers like O_3 or, especially, O_4 and the polynomial. To minimize this problem the polynomial degree should be chosen as low as possible. Obviously, two (or more) cross sections of the same absorber at different temperatures are quite similar in structure and, thus, strongly correlate. This can be avoided by a mathematical orthogonalization procedure which makes the cross sections linearly independent. Generally, correlations increase with the degrees of freedom of the fit (number of molecules included in the fit N and polynomial degree M) and decrease with the number of entries in the measurement vector y (fit range, i.e. numbers of pixels).

Corrections and shortcomings of the DOAS forward model

In the following some effects, not considered by the DOAS model function, equation (4.16), are discussed and, if possible, a correction is implemented in the spectral retrieval.

Solar I_0 effect
The condition for the linearisation of Beer-Lambert's law of a constant reference spectrum $I_{ref,b}$ is generally not satisfied (see above). I_{ref} exhibits many Fraunhofer absorption lines which cause a highly structured spectral pattern. Since the DOAS model function is constructed from the linearized version of Beer-Lambert's law (equation (4.15)) it cannot perfectly model the measured spectrum. The systematic deficiency is referred to as solar I_0 effect. Qualitatively spoken, it arises from the exchange of convolving and taking the logarithm of the recorded radiance. The solar I_0 effect is of importance when strong absorbers interfere with underlying weakly absorbing species. It can be corrected by attributing

4.1. SPECTROSCOPIC ANALYSIS

the systematic residual structures originating from non-constant I_{ref} to the molecular absorption cross sections $\sigma_{m,d}$ (Huppert, 2000). The slant column density SCD_i has to be guessed a priori, then the corresponding I_0-corrected molecular absorption cross section $\sigma_{i,d,corr}(\lambda)$ is given by

$$\sigma_{i,d,corr}(\lambda) = \frac{1}{SCD_i} \ln \frac{\left(I_0(\lambda) e^{-\sigma_{i,d}(\lambda) SCD_i}\right) \otimes g(\lambda)}{I_0(\lambda) \otimes g(\lambda)}. \tag{4.19}$$

where $\sigma_{i,d}(\lambda)$ is the high resolution absorption cross section, $I_{ref}(\lambda)$ a high resolution solar spectrum, e. g. Kurucz et al. (1984), and $g(\lambda)$ the instrument function of the spectrograph.

Instrumental stray light
Second or higher orders of the grating reflecting light, or light of wavelengths outside the detection range lead to spectrometer stray light. This causes an additive offset to the measured intensity and, thus, changes the optical density of the Fraunhofer lines or the molecular absorption. One possibility to reduce this effect is the use of filters cutting off the undetected wavelengths. The change in stray light contribution can be corrected for during the fitting process by introducing a polynomial $O(n)$ of up to 2nd order in the fitting process as follows

$$\chi^2 = \frac{1}{W - V - 1} \sum_{n=1}^{W} \left(\frac{\ln\left[I^*(n) - O(n)\right] - F(n)}{\epsilon_n} \right)^2. \tag{4.20}$$

The polynomial coefficients are additional non-linear parameters which are determined by the Levenberg-Marquardt fitting algorithm.

Instrumental shortcomings

A stable optical imaging is very important for any spectroscopic analysis. A change in ambient pressure and temperature can lead to a change of the resolution and/or the pixel-wavelength mapping due to mechanical relaxation and change of the refractive index of the air inside the spectrometer. As a possible shift is included in the forward model, the shift in wavelength space is accounted for. But still this effect leads to additional residual structures because of a varying sensitivity of the individual pixels of the CCD array. In order to obtain the sensitivity of the pixels experimentally, the unstructured spectrum of a halogen lamp is recorded. By dividing the measured spectra by the low pass filtered lamp spectrum, the effect of the non-constant diode sensitivity in case of a shift is reduced (Weidner, 2005).
Structures of one spectrum also appearing in the next one are called "memory effect" and can be avoided by performing a dummy readout with the lowest possible integration time after each change in viewing geometry and integration time.

Temperature and pressure dependence of the absorption cross sections
Shape and absolute value of the UV/vis absorption cross sections are strongly dependent on temperature and, to a lesser degree, on pressure. On one hand this leads to inappropriate fit parameters a_j for different temperatures due to a change of the absolute value. On the other hand it does not allow to perform an adequate model of the measured radiance and can result in large residual structures. Hence, if the temperature dependence of the cross section of a strong absorber is not taken into account, the detection of underlying weak absorbers may be impossible. The latter effect can be minimized

by including in the DOAS fit cross sections measured at different temperatures. If the temperature dependence is approximately linear, two cross sections are sufficient to cover a large temperature region, i.e. the region where the temperature dependence is linear. In order to avoid correlations between the two cross sections at different temperatures, one of them is orthogonalized with respect to the other. Therefore, their differential structure is obtained by fitting a polynomial. Note that the fit parameter of the orthogonalized cross section can not be interpreted as a slant column density. The difference of the absolute size of the cross sections at different temperatures can be corrected by a procedure described in Butz et al. (2005). In this thesis there are only two cross sections at different temperatures fitted in order to remove their absorption structure to obtain the slant column density of another gas. If the SCD of the same gas is the desired quantity, only one cross section is fitted.

4.1.3 DOAS analysis of the particular gases

The spectral analysis for this work is performed by the fitting routine of WinDOAS (Fayt and van Roozendael, 2001), developed at BIRA/IASB. WinDOAS is widely used and tested, hence it has become a standard in DOAS evaluation. Preprocessing of the spectra, convolution of the high resolution cross sections and calculation of the Ring spectrum (Grainger and Ring, 1962) is performed by the Windows DOAS tool DOASIS, developed at the IUP Heidelberg by (Kraus, 2004b). This includes the correction for offset and dark current and the time weighted division by a high pass filtered halogen lamp to eliminate the pixel-to-pixel structures.

The high resolution cross sections are convoluted to the instrumental resolution with the instrumental slit function determined from a recorded line spectra of a HgCd lamp. The line at $\lambda = 468$ nm is used for the evaluations in the visible range and $\lambda = 360$ nm for the evaluations in the UV. As the Fraunhofer reference, a spectrum with low absorption of the particular trace gas absorption is chosen and the correction spectrum for the Ring effect (Grainger and Ring, 1962) is calculated.

In the following a detailed description of the DOAS evaluation of the particular gases is given, following the recommendations given in (Weidner, 2005). The cross sections are listed in Table 4.1.3.

Table 4.1: Compendium of molecular absorption cross sections used for the spectral analysis.

Species	Reference	T_1	T_2
O_3	Voigt et al. (2001)	223 K	203 K
NO_2	Harder et al. (1997)	230 K	217 K
H_2O	Rothman et al. (2005)	230 K	-
O_4	Hermans (2002)	room temp.	-
BrO	Wahner et al. (1988)	228 K	-
HONO	Stutz et al. (2000)	294 K	-

4.1. SPECTROSCOPIC ANALYSIS

O_3

The spectral retrieval of O_3 (see Figure 4.2) is performed in the 490 − 520 nm wavelength range and includes the cross sections of O_3 at T_1, corrected for I_0 effect, O_4, NO_2 at T_1 and T_2 (with T_1 orthogonalized with respect to T_2) and H_2O, both corrected for I_0 effect. A polynomial of 4th degree is used to account for the broad band structures and an additive polynomial of 2nd degree is included to account for stray light in the spectrograph. A spectrum correcting for the Ring effect (Grainger and Ring, 1962) is also included in the fitting routine (Bussemer, 1993).

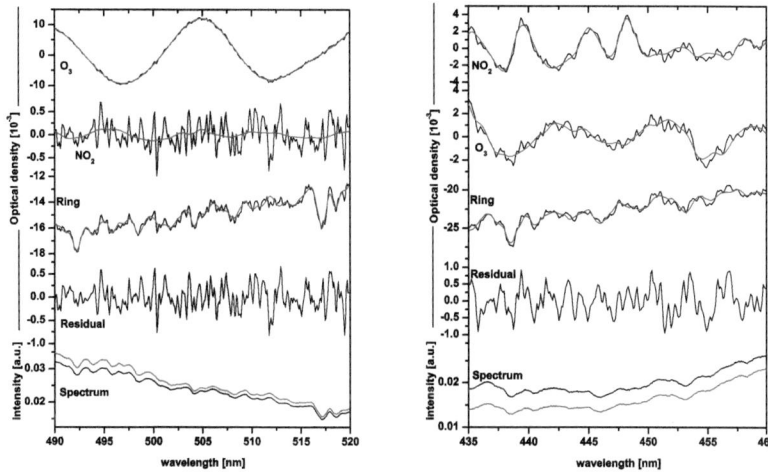

Figure 4.2: Sample DOAS evaluation of O_3 (left panel), which is performed in the wavelength interval 490−520 nm and of NO_2 (right panel), which is performed the in the wavelength interval 435−460 nm. Shown is the optical density of the absorbance of the trace gases and the Ring effect. The lowest two traces show the measured (red line) and the reference (black line) spectra. The panels above illustrate the remaining residuals of the fitting procedure. The red lines indicate the spectral absorption and the black lines the sum of the spectral absorption and the residual.

NO_2

The spectral retrieval of NO_2 (see Figure 4.2) is performed in the 435 − 460 nm wavelength range and includes the cross sections of NO_2 at T_1 corrected for I_0 effect, O_3 at T_1 and T_2 (with T_1 orthogonalized with respect to T_2), both corrected for I_0 effect, O_4 and H_2O. A polynomial of 4th degree is used to account for the broad band structures and an additive polynomial of 2nd degree is included to account for stray light in the spectrograph. A spectrum correcting for the Ring effect is also included in the fitting routine.

O_4

The spectral retrieval of O_4 (see Figure 4.3) is performed in the 465 − 490 nm wavelength range and includes the cross sections of O_3 at T_1 and T_2, both corrected for I_0 effect, NO_2 at T_1 and T_2 (with T_1 orthogonalized with respect to T_2), both corrected for I_0 effect and O_4. A polynomial of 3rd degree is used to account for the broad band structures but no additive polynomial is fitted to minimize correlations as its effect is rather small. A spectrum correcting for the Ring effect is also included in the fitting routine.

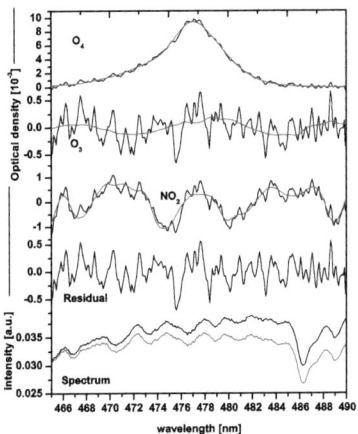

Figure 4.3: Sample DOAS evaluation of O_4, which is performed in the wavelength interval 465−490 nm. Shown is the optical density of the absorbance of the trace gases and Ring. The lowest two traces show the measured (red line) and the reference (black line) spectra. The panels above illustrate the remaining residuals of the fitting procedure. The red lines indicate the spectral absorption and the black lines the sum of the spectral absorption and the residual.

BrO

The spectral retrieval of BrO (see Figure 4.4) is performed as recommended in Alliwell et al. (2002), in the 346 − 359 nm wavelength range. It includes the cross sections of BrO shifted by +0.25nm to match the wavelength calibration from the IUP Bremen Fleischmann et al. (2000), O_3 at T_1 and T_2 (with T_1 orthogonalized with respect to T_2) both corrected for I_0 effect, NO_2 at T_1 corrected for I_0 effect and O_4.

4.1. SPECTROSCOPIC ANALYSIS

A polynomial of 2nd degree is used to account for the broad band structures and an additive polynomial of 2nd degree is included to account for stray light in the spectrograph. In general the detection of BrO is more challenging compared to NO_2 and O_3. The amount of light in the UV reaching the spectrograph is, depending on viewing geometry less (as compared to the visible spectral range) and therefore the noise level is usually higher. In addition the optical densities are much lower compared to the other gases. The much higher O_3 absorption has to be completely removed, as remaining O_3 structures might be mistaken as BrO.

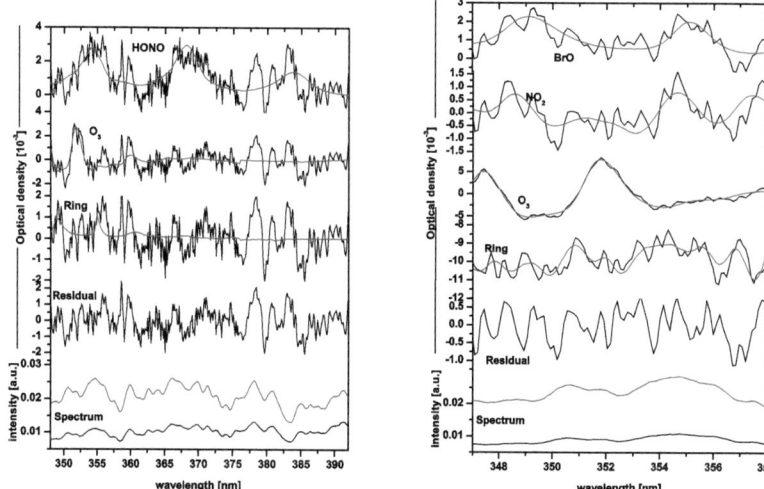

Figure 4.4: Sample DOAS evaluation of HONO (left panel), which is performed in the wavelength interval 348 − 392 nm and of BrO (right panel), which is performed in the wavelength interval 346 − 359 nm. Shown is the optical density of the absorbance of the trace gases and the Ring. The lowest two traces show the measured (red line) and the reference (black line) spectra. The panels above illustrate the remaining residuals of the fitting procedure. The red lines indicate the spectral absorption and the black lines the sum of the spectral absorption and the residual.

HONO

The spectral retrieval of HONO (see Figure 4.4) is performed in the 348 − 392 nm wavelength range. It includes the cross sections of HONO, BrO (shifted, as described above), O_3 at T_1 and T_2 (with T_1 orthogonalized with respect to T_2) both corrected for I_0 effect, NO_2 at T_1 corrected for I_0 effect and O_4.

A polynomial of 2nd degree is used to account for the broad band structures and an additive polynomial of 2nd degree is included to account for stray light in the spectrograph.

4.2 Radiative transfer modeling

The interpretation of the ΔSCDs (as output of the spectral analysis), in terms of concentration requires knowledge about the line-of-sight, radiative transfer is modeled.
During the course of this work three radiative transfer models were used, that have been developed subsequently at the Institute of Environmental Physics in Heidelberg: Tracy, TracyII and McArtim. For the final interpretation of the measurements, the latest of them, McArtim is applied.

4.2.1 RTM McArtim

McArtim is an acronym for Monte Carlo Atmospheric Radiative Transfer Inversion Model. It is a fully spherical 3-D RTM that has been designed to simulate the radiative transfer in the atmosphere of the Earth in the UV/vis/NIR spectral range with Monte Carlo methods. McArtim generates an ensemble of individual photon path trajectories through the simulated atmosphere, each representing a solution of the adjoint monochromatic radiative transfer equation. The RTM has been written at the Institute of Environmental Physics, University of Heidelberg and validated with other radiative transfer models (Wagner et al., 2007) and measurements (Deutschmann, 2008). The advantage of the Monte Carlo approach is its flexibility. All scattering and attenuation processes relevant to the radiative transfer are implemented. Ground albedo, multiple trace gas concentrations and aerosol load profiles, including various aerosol scattering parameterizations can be chosen by the user. For this work, a model atmosphere from 0 to 70 km, discretized in layers of 1 km altitude is used, where properties like air density, humidity, aerosol and temperature are prescribed. The viewing geometry is defined by different detector locations (latitude, longitude, and altitude) and variable viewing directions (elevation α and azimuth angle) of the detector with variable field of view (e.g. rectangular or rectangular gaussian).
Among many other parameters, the program computes the so-called Box Air Mass Factors (BoxAMF), i.e. discretized weighting functions, which are defined as the ratio of the effectively traveled light path inside a given box and its height. Figure 4.5 displays modeled BoxAMFs for one sequence of elevation angles as they are performed during balloon float. The weighting functions from the Limb scanning measurements show pronounced maxima at heights that strongly depend on the elevation angle. E.g. for $\alpha = -2.25°$, the tangent height (i.e. altitude layer where the BoxAMF gets maximal) is about 30 km and for $\alpha = -4.25°$ about 16 km, both for an balloon float altitude of around 34 km. For $\alpha \geq -2.25°$, the tangent height coincides with the detector altitude but the BoxAMF, especially at the detector altitude, is strongly decreasing with increasing elevation angle. The BoxAMFs are used to calculate the SCDs of the trace gases by multiplying them with concentration and box height and summing up over all boxes. These simulated SCDs are directly compared to the measurements.
Another output is the radiance in the direction of detection which is compared to the measured intensity in order to test the accuracy of the assumed viewing geometry, as described in the following section. In the retrieval of trace gas profiles McArtim provides the weighting function in altitude space.

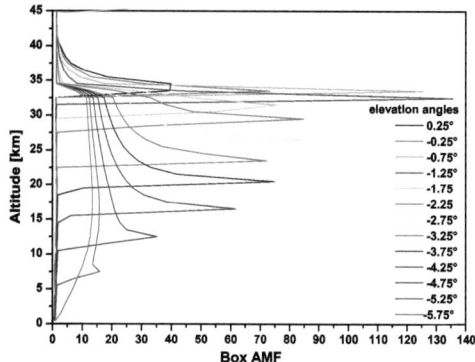

Figure 4.5: BoxAMFs ($K_{i,j}$) for the NO_2 concentration profile, for Limb scans, recorded at 35 km altitude in series from 0.5° to -5.5° elevation angle from aboard the LPMA/IASI payload on June 30, 2005.

4.3 Retrieval of the elevation α

Several viewing geometry parameters are necessary to built the forward model for the retrieval of trace gas profiles. They are referred to as forward model parameter, as they have an influence on the forward model, but are not going to be retrieved. As mentioned above, the SZA and SRAA are calculated using a record of the gondola's geolocation and attitude. As a result, in relation to their effect (see Weidner (2005)) they are well known. However, the elevation angle α of the telescope, which has strong influence on the observed quantities (see weighting functions 4.5), is also affected by pendulum oscillations $\Delta\alpha^{**}$ of the gondola and due to possible primary calibration errors subject to uncertainty (section 3.2.3). Primal calibration errors and oscillations are treated separately, since the first, due to its systematic nature, can be retrieved from measured intensities and the latter is more random and just characterized as an error. For tracing the line-of-sight from the sun to the telescope, the viewing geometry has to be defined. Observation height and the azimuth angle are obtained from the attitude control systems of the individual payloads. The actual elevation angle of the telescope relative to the payload orientation is controlled by the mini-DOAS instrument itself. As the relative position of the telescope to the gondola α^* affects the absolute elevation angle α, it is carefully aligned to the principle axis of the payload prior to the balloon flight (see section 3.2.3).

A remaining misalignment can be tested after the flight by comparing the modelled and measured relative radiances for each observation. This procedure serves as a sensitivity test for pointing, since the radiance of skylight in the UV/vis spectral range near the horizon changes largely with tangent height and shows a wavelength depend maximum in the lowermost stratosphere (Sioris et al., 2004; Weidner, 2005). In contrast to tropospheric measurements, unknown optical constituents such as aerosols and clouds barely

influence the radiance distribution (with the exception of periods after a large scale volcanic eruption). The resulting viewing geometry then acts as the final input for RTM.
A possible contribution of stray light to the measured signal is estimated by Weidner (2005) to 0.2% at the low (around 350 nm) and 0.3% at the high wavelength (around 500 nm) end of the detector for full saturation. This means that even for poorly saturated spectra, e.g. in the UV at tropospheric altitudes where the saturation level is only around 20%, the atmospheric straylight ratio is around 1%.
The shape of the measured and modeled intensities can be matched for each wavelength λ in order

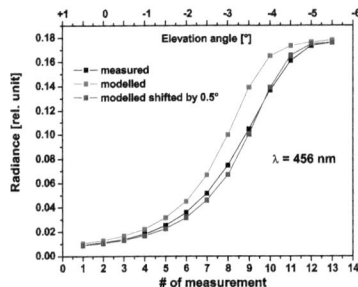 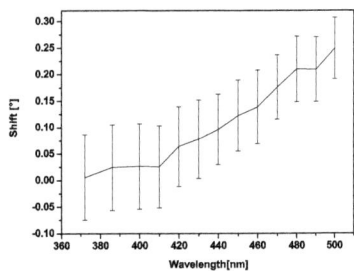

Figure 4.6: Left panel: Modeled (red) and measured (black) Limb radiances, with the modeled values shifted to measured values (blue). Right panel: Wavelength dependency of the retrieved $\Delta\alpha$.

to retrieve the elevation angle offset $\Delta\alpha$ as a shift, and the transmittance of the detector as a fitting coefficient t.

$$I_{mod}(\alpha, \lambda) = t(\lambda)[I_{mess}(\alpha + \Delta\alpha)] \qquad (4.21)$$

where I_{mod} represents the modeled intensity (radiance), α the elevation angle, t the transmittance of the detector, I_{mess} the intensity (number of counts at a certain wavelength) and $\Delta\alpha$ the shift of the elevation angle.
In fact, there is a wavelength dependency of the retrieved $\Delta\alpha$ (see Figure 4.6), such that the retrieved elevation α for lower wavelengths is lower than for higher wavelengths, which is due to the wavelength dependent effect of the gondola oscillation on the line-of-sight (explained in the following).

4.3. RETRIEVAL OF THE ELEVATION α

High frequency oscillations of the gondola

High frequency oscillations of the gondola can be accounted for in the RTM calculations by adjusting the effective field of view. A permanent high frequency pendulum oscillation in elevation angle of $\Delta\alpha^{**} \geq 0.1°$ can be concluded from the available records of the attitude control systems.
In practice, the field of view is thus expanded to a Gaussian shape with the variance $\sigma=0.3°$. The effect on the line-of-sight is illustrated in Figure 4.7 and shows a broadening of the modelled box airmass factors and therefore a decrease in altitude resolution.

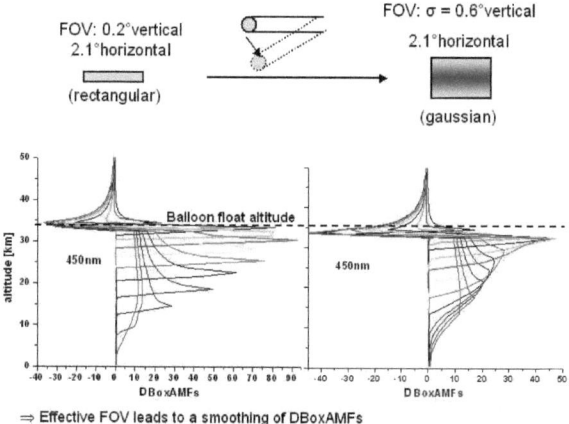

⇒ Effective FOV leads to a smoothing of DBoxAMFs

Figure 4.7: Effect of a Gaussiantype FOV on the ΔBoxAMFs. Left panel: Rectangular instrumental FOV and corresponding BoxAMFs. Right panel: Gaussiantype FOV for a variance $\sigma = 0.3°$ and corresponding BoxAMFs.

How remaining and unaccounted pointing errors due to gondola oscillation propagate into the results is discussed in section 4.3. The remaining error in the elevation α propagates as forward parameter error into the retrieval of trace gas profiles as described in Section 4.4.5.

4.4 Retrieval of trace gas profiles

In the previous chapter the retrieval of ΔSCDs from measured wavelength dependent intensities has been characterized. These ΔSCDs still include attributes of both, the atmosphere and the measurement method, as they depend on the observation geometry of the individual measurement. However, for the comparison with other measurements using different methods and in order to draw final conclusions concerning photochemistry, the desired product needs to become attributed to the atmosphere alone. Hence, the goal of our spectroscopic measurements are concentration vertical profiles, i. e. the particular trace gas concentration as a function of altitude. In analogy to equation 4.1, a model F can be setup which links the measurements $\mathbf{y}$ (ΔSCDs) to the retrieval parameters $\mathbf{x_j}$ and forward model parameters $\mathbf{b_j}$

$$\mathbf{y_i} = \mathbf{F}(\mathbf{x_j}, \mathbf{b_j}) + \epsilon. \tag{4.22}$$

Here, in contrast to section 4.1, the measurement vector $\mathbf{y}$ is represented by an ensemble of Δ SCDs and the retrieval vector component $\mathbf{x_j}$ is the concentration of the considered trace gas on the chosen altitude levels j.

As long as measured optical thickness is much smaller than unity, the kernel $\mathbf{K_{i,j}}$ can be approximated linearly and equation 4.22 can be written in linear form

$$\mathbf{y_i} = \mathbf{K_{i,j}}(\mathbf{b_j})\,\mathbf{x_j} + \epsilon. \tag{4.23}$$

In the framework of this thesis a new method for the retrieval of temporally subsequent concentration profiles from Limb scanning measurements is developed, where the contribution of a measurement to the retrieved profile is weighted by the inverse of the time lag between both. This time weighting becomes important if the desired concentration changes with time.

For the retrieval of trace gas profiles from balloon-borne mini-DOAS measurements so far no time weighting was implemented. Butz (2006) implemented a retrieval for direct sun measurements taking into account the change of SZA along the line-of-sight, using a chemical model to calculate photochemical weighting factors. E.g. Schofield et al. (2004) and Hendrick et al. (2004a) also employed a chemical model to calculate the change in concentration along the SZA. As in the case of our scattered light measurements the change in SZA along the line-of-sight is less than $1.21°$ for all viewing geometries and the relative change of the measured gases along $1.21°$ ΔSZA does not exceed 1% during daytime, it is possible within a given limit of error to assume a constant concentration at a certain altitude along the line-of-sight. This assumption renders photochemical modelling in the forward model unnecessary and allows us to separate the time and the geometrical weighting.

The desired state of the time dependent retrieval consists of the elements $x_{j,k}$, which are the absorber concentrations at the atmospheric layer j for temporary subsequent profiles k. Hence we define a kernel (weighting function $\mathbf{K}$), which involves both, time and space weighting:

$$y_i = \mathbf{K_{i,j,k}} \cdot x_{j,k} + \epsilon \tag{4.24}$$

Note, that the indices i and k contain implicitly the time t_i at which the measurement is taken and the time T_k for which the profile k is inferred.

The geometrical weighting is inferred from RTM calculations and is represented by the matrix $\mathbf{L_{i,j}}$.

4.4. RETRIEVAL OF TRACE GAS PROFILES

Accordingly $\mathbf{L_{1,J}}$ gives the sensitivity of the measurement SCD_i to the absorber concentration x_j, neglecting the time difference between T_k and t_i.

$$\mathbf{L_{1,J}} = \frac{dSCD_i}{dx_j} \quad (4.25)$$

Finally the time weighting matrix $\mathbf{C_{1,k}}$ represents the sensitivity of the measurement SCD_i at time t_i to the absorber concentration x_k at time T_k. The product of the two matrices $\mathbf{L}$ and $\mathbf{C}$ gives the Kernel considering both, geometrical and time weighting.

4.4.1 Time weighting

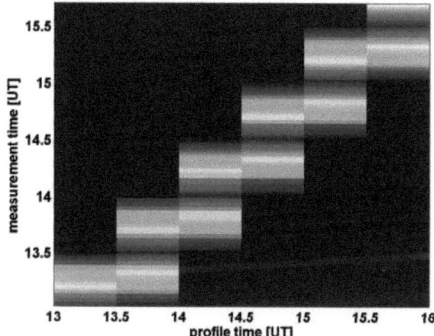

Figure 4.8: Example time weighting matrix C for the Limb observation from aboard the MIPAS-B payload over Teresina on June 14, 2005.

The state $x_{j,k}$ includes concentration profiles at subsequent times T_k, which are defined by the time grid of the kernel and state vector. Prior to the retrieval, the time grid is defined utilizing the period for which measurements are available. On one hand, as the retrieval of one profile requires several measurements y_i, the time grid of the measurement vector can not be the same as for the state vector, but has to be somewhat finer. On the other hand, if the diurnal variation of a certain trace gas is the required state, the time interval between two profiles should still be short enough to reasonably represent the gradient in concentration over time. For the comparison with measurements performed on different platforms, the time grid is chosen to match the particular time of the profile to be compared with. For the transformation between the measurement grid and the state grid, $\mathbf{C_{1,k}}$ is defined, which characterizes the time-lag between the actual measurement and the state. It quantifies the contribution of each measurement to the two nearest states, adding weights together to unity. $\mathbf{C_{1,k}}$ is derived from the relation of time differences between measurement and state, where T_k denotes the fixed times of the state:

for $T_k \leq t_i \leq T_{k+1}$:

$$\mathbf{C_{i,k}} = \frac{T_{k+1} - t_i}{T_{k+1} - T_k} \qquad (4.26)$$

$$\mathbf{C_{i,k+1}} = \frac{t_i - T_k}{T_{k+1} - T_k} = \mathbf{C_{i,k}} - 1 \qquad (4.27)$$

and for $t_i \leq T_k$ or $T_{k+1} \leq t_i$ $\mathbf{C_{i,k}} = 0$

4.4.2 The combined kernel

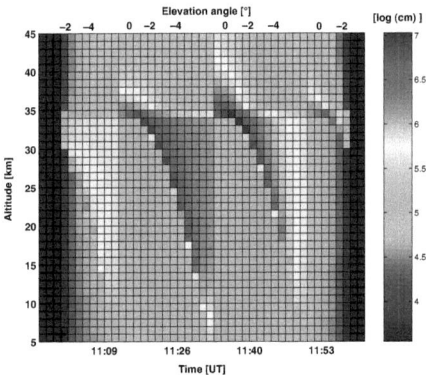

Figure 4.9: Logarithm of the kernel $\mathbf{K}_{i,j}$ for the retrieved NO_2 concentration profile at $T_k = 11:30$ [UT]. The example is for Limb scanning measurements at 35 km altitude with subsequent measurements in steps of $0.5°$ from $\alpha = 0.5°$ to $\alpha = -5.5°$ elevation angle from aboard the LPMA/IASI payload on June 30, 2005.

The combined weighting function matrix representing the sensitivity of a measurement to the state (absorber concentration profile at the times $\mathbf{T}_k$) is calculated from

$$\mathbf{\widetilde{K}_{i,j,k}} = \mathbf{L_{i,j} C_{i,k}} \qquad (4.28)$$

The differential character of the measurements is considered by inserting Δ SCD_i from equation 4.6 into equation 4.24

$$y_i := \Delta SCD_i = SCD_i - SCD_{ref} \qquad (4.29)$$

$$= \mathbf{\widetilde{K}_{i,j,k}} x_{j,k} - \mathbf{\widetilde{K}_{ref\,j,k}} x_{j,k} = (\mathbf{\widetilde{K}_{i,j,k}} - \mathbf{\widetilde{K}_{ref\,j,k}}) x_{j,k} \qquad (4.30)$$

and thus the final kernel is defined as

$$\mathbf{K_{i,j,k}} := \mathbf{\widetilde{K}_{i,j,k}} - \mathbf{\widetilde{K}_{ref\,j,k}} \qquad (4.31)$$

4.4. RETRIEVAL OF TRACE GAS PROFILES

by which the weight of the reference measurement is by definition zero. In linear approximation equation 4.24 can be solved in one iteration as shown in section 4.4.4.

As an example, Figure 4.9 shows the contributions of the Limb scanning measurements at 35 km between t_i = 10:42 UT and t_i=12 UT to the inferred profile at T_k = 11:30 UT. The contribution of the measurements before 11 and after 12 UT to the inferred profile (11:30 UT) is zero. Concerning the weighting in altitude space the impact of the varying elevation angle α is clearly visible; horizontal elevation α has the highest sensitivity at float altitude, and the lowest elevation α has the highest sensitivity for the results at lower altitudes.

The sensitivity of the measurement y_i to the overhead trace gas column is very low, as the light path through the atmosphere above the gondola is nearly equal for all viewing geometries (namely the vertical path), hence the difference of $\widetilde{\mathbf{K}}_{i,j,k}$ and $\widetilde{\mathbf{K}}_{ref,j,k}$, meaning $\mathbf{K}_{i,j,k}$ is close to zero for j ranging between 40 and 70 km.

In order to proof if the forward model is able to explain the measurements, a comparison of measured and forward modeled $\Delta SCDs$ is shown in Figure 4.10.

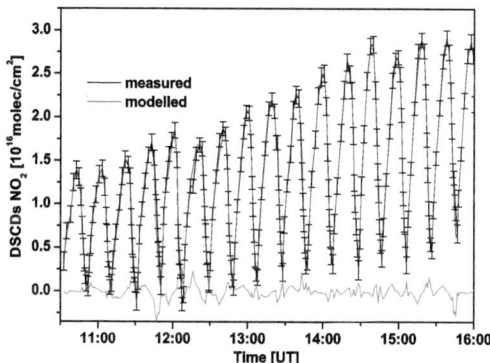

Figure 4.10: Measured (black) and forward modeled (red) $\Delta SCDs$ for the observation from aboard LPMA/IASI payload on June 30, 2005. The difference between modeled and measured $\Delta SCDs$ is shown in green.

4.4.3 Inverse method - optimal estimation

After finding a kernel which is able to explain the measurements, the equation 4.22 is to be solved for x. The retrieval of trace gas profiles from balloon-borne Limb scattered skylight measurements is formally ill-posed. An equivalent description of this fact is that the matrix $\mathbf{K}^T \mathbf{S}_\epsilon^{-1} \mathbf{K}$ is close to zero. Several conditions have to be fulfilled to allow a matrix to be inverted, which is like a system of linear equations, that has to be solved. The matrix has to be square, in terms of the system of equations this means that, the number of equations has to be equal to the number of unknowns. Also the rows of the matrix or each equation has to be linearly independent. In the case of the Limb scanning measurements, measurements of different viewing directions are easy to distinguish for heights below float altitude, but not above. However, in order to invert the matrix and to obtain a physically reasonable solution from the manifold of all mathematically possible solutions, some kind of a prior constraint needs to be employed to make the problem well posed. This can be achieved by estimating the null-space and near-null space components (i.e. the concentration of a trace gas above 40 km) of the retrieval via a priori information like in the maximum a posteriori method or by elimination of the null space with an appropriate retrieval grid (see Section 4.4.4). The matrix to be inverted then reads $\mathbf{K}^T \mathbf{S}_\epsilon^{-1} \mathbf{K} + \mathbf{H}$ where $\mathbf{H}$ is the additional constraint. In this thesis mainly an optimal estimation method (Rodgers, 2000) is applied, as it is common practice in this field (e.g.Schofield et al. (2004); Hendrick et al. (2004a); Weidner (2005)). Additionally a second method is compared to the optimal estimation method(see section 4.4.4).

The optimal estimation method uses prior knowledge in terms of a priori probability, where the *a priori* x_a is a climatological expectation of the quantity, to retrieve, and $\mathbf{S_a}$ is its covariance. Then the retrieved state is a result of both, the a priori profile x_a and the inverted measurements, each weighted by its covariance $\mathbf{S_a}$ and $\mathbf{S}_\epsilon$ respectively (Rodgers, 2000). We set up a cost function $\chi^2(x)$, which is a measure of the deviation of the simulated measurement from the real measurement

$$\chi^2(x) = [K_{i,j,k}\hat{x}_{j,k} - \Delta SCD_i]S_{\epsilon,i}[K_{i,j,k}\hat{x}_{j,k} - \Delta SCD_i] + [\hat{x}_{j,k} - x_{a,j,k}]S_{a,j,k}[\hat{x}_{j,k} - x_{a,j,k}] \quad (4.32)$$

The measurement covariance $S_{\epsilon,ii}$ and the a priori covariance $S_{a,j,k}$ matrices contain the squares of the standard deviations $\sigma_{\epsilon,i}$ and $\sigma_{a,j,k}$ on the diagonal. In the case of the measurement covariance these are the errors of the ΔSCDs, which are given as an output of the retrieval software winDOAS. Non diagonal elements of $\sigma_{\epsilon,i}$ are zero since cross correlations of the DOAS analysis are neglected. Actually, the optimal estimation method requires an a priori set including a known covariance. Since the covariance of the a priori $\mathbf{S_a}$ is used here as a tuning parameter, the retrieval is not optimal in its original sense. The entries in the diagonal of the a priori covariance matrix are a squared percentage p of the a priori value.

$$\mathbf{S_a}_{j,j} = \frac{p}{100} x_{aj} \quad (4.33)$$

where the percentage p is determined by an L-curve method. The goal is to retrieve most of the information from the measurements, without over interpreting them (i.e. mapping measurement errors into the state). A low error in the a priori x_a results in a retrieval highly relying on this a priori with the information from the measurements being largely ignored. On the other hand a large error in the construction of $\mathbf{S_a}$ results in the a priori being ignored and the retrieval being an unrealistic over-interpretation of the measurements. In order to find an appropriate $\mathbf{S_a}$, p is varied from a very low value (tightly constrained) to a very high value (loosely constrained). The root mean square of the residual of the fit is plotted versus p is constructed (see Figure 4.11). The chosen p is the value where increasing the a priori error no longer results in a marked improvement of the measurement fit. It ranges usually between 40% and 80%.

There exists a class of inversion methods which fill the extra-diagonal elements of $\mathbf{S_a}$ with non-zero values in order to get a smooth result (regularisation). Within the framework this thesis $\mathbf{S_a}$ also contains extra-diagonal terms in order to account for correlations between concentrations at different altitude levels. These terms are added as Gaussian functions as follows

$$\mathbf{S_a}_{i,j} = \sqrt{(\mathbf{S_a}_{i,i} \cdot \mathbf{S_a}_{j,j} \cdot \exp(-log_2((i-j)/h)^2))} \quad (4.34)$$

4.4. RETRIEVAL OF TRACE GAS PROFILES

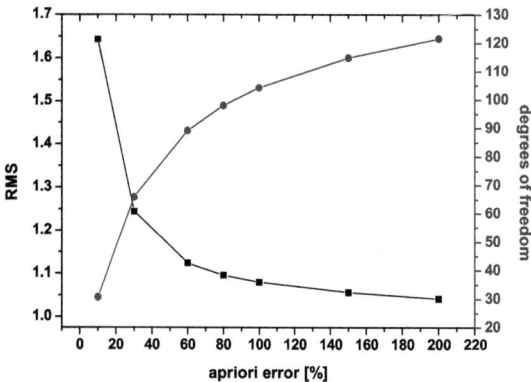

Figure 4.11: L-curve estimation of the a priori error S_a.

with half width at half maximum (HWHM) of the Gaussian function, h (or length scale h (Rodgers, 2000)). Sensitivity runs with varying h were performed (Hendrick et al., 2004a), in order to maximize the number of degrees of freedom of the retrieval. A maximum is found for h = 0.5 (see Figure 4.12, standing for a correlation length of 1 km. This finding is in agreement with the height resolution indicated by the width of the averaging kernels (see chapter 5).

The impact of this correlation length on the degrees of freedom of the retrieval can be seen in Figure 4.12 where it is plotted as a function of the parameter h, which is the half of the correlation length (see Eq. 4.4.3. The trace of A is maximum (85) for a correlation length of 1 km (h =0.5 km), which is the choice of all retrievals shown in this work.

Using non diagonal elements in the a priori covariance provides a link between different entries in the state vector, i.e. in our case different altitudes. Consequently this is a constraint on the smoothness of the profile which is similar to the constraints given by regularization methods (Rodgers, 2000).

Referring to equation 4.32, the inverse problem is then a $\chi^2(x)$-optimization problem, i.e. the solution is a state $\hat{x}$ that minimizes the cost function has to be found. If the a priori profile is considered, the resulting state is the best (optimal) compromise between the a priori knowledge and the measurement. The criterion which $\hat{x}_{j,k}$ has to fulfill, in order to minimize the cost function, is to fit the simulated $K_{i,j,k}\hat{x}_{j,k}$ to the measurement SCD_i is

$$\nabla \chi^2(x) = 0 \qquad (4.35)$$

$$\nabla K_{i,j,k}\hat{x}_{j,k} S_{\epsilon,i}[K_{i,j,k}\hat{x}_{j,k} - \Delta SCD_i] + S_{a,j,k}[\hat{x}_{j,k} - x_{a,j,k}] = 0 \qquad (4.36)$$

In the case of a linear forward function (approximated by a linear forward model), we can write

$$\nabla K_{i,j,k}\hat{x}_{j,k} = K_{i,j,k} \qquad (4.37)$$

The minimization problem then is:

$$\nabla K_{i,j,k} S_{\epsilon,i}[K_{i,j,k}\hat{x}_{j,k} - \Delta SCD_i] + S_{a,j,k}[\hat{x}_{j,k} - x_{a,j,k}] = 0 \qquad (4.38)$$

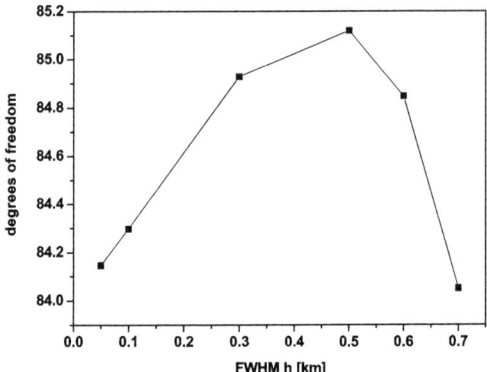

Figure 4.12: Degrees of freedom of the retrieval (trace of the averaging kernels matrix A) plotted as a function of the HWHM (h). This curve has been calculated for the retrieval of NO_2 from Limb scanning on June 30, 2005.

which is solved for
$$\hat{x} = (\mathbf{K}^T \mathbf{S}_\epsilon^{-1} \mathbf{K} + \mathbf{S_a}^{-1})^{-1}(\mathbf{K}^T \mathbf{S}_\epsilon^{-1} y + \mathbf{S_a}^{-1} x_a) \qquad (4.39)$$
where the superscript T indicates transposed matrices.

4.4.4 Other methods to constrain the retrieval

The method chosen here to constrain the retrieval is, as already mentioned, not a pure optimal estimation retrieval since the a priori covariance is not known and there are additional non diagonal elements in the a priori covariance. However, other methods to constrain the retrieval were tested and are briefly explained here.

Tikhonov regularization

This method, as used here, minimizes the deviation of the solution from the measurements as well as the mean squared second departure from zero (which is derived by multiplying with the second derivative operator L, explained below). This gives a smoothing constraint to the solution and is helpful, as there is no a priori information. A factor γ is chosen to give appropriate relative weighting to the two constraints (Rodgers, 2000). The solution is then given by
$$\hat{x} = (\mathbf{K}^T S_\epsilon^{-1} \cdot \mathbf{K} + \gamma^{-1} \cdot \mathbf{H})^{-1} \cdot (\mathbf{K}^T \cdot S_\epsilon^{-1} \cdot y) \qquad (4.40)$$
where $\mathbf{H}$ is the smoothing constraint $\mathbf{H} = \alpha \mathbf{L}^T \mathbf{L}$ and $\mathbf{L}$ is the second derivative operator

4.4. RETRIEVAL OF TRACE GAS PROFILES

Table 4.2: The second derivative operator L.

$$\begin{array}{ccccccc} 1 & -2 & 1 & 0 & 0 & 0 & 0 \\ 0 & 1 & -2 & 1 & 0 & 0 & 0 \\ 0 & 0 & 1 & -2 & 1 & 0 & 0 \\ 0 & 0 & 0 & 1 & -2 & 1 & 0 \\ 0 & 0 & 0 & 0 & 1 & -2 & 1 \end{array}$$

This is the same as the MAP solution with the inverse of H as a priori covariance and the *a priori* itself set to zero. It also bears some resemblance to the truncated singular vector decomposition method with a choice of eigenvalues determined by gamma.

Constraining by a coarse retrieval grid

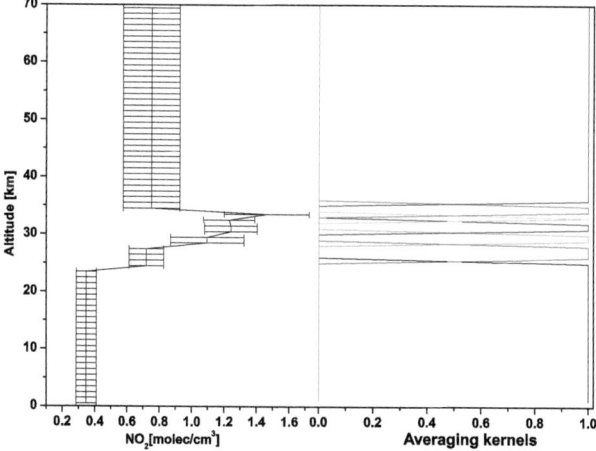

Figure 4.13: Left panel: Concentration of NO_2 retrieved on a coarse retrieval grid. Right panel: Averaging kernels from the retrieval on a coarse grid.

The state vector does not need to consist of the number of unknowns on the forward model grid. In fact the number of unknowns can be reduced by a transformation to a coarser grid by a matrix textbfW (Rodgers, 2000). In order to be independent of any a priori constraint (such as having knowledge on the quantity to be retrieved prior to the measurement) the grid should be chosen such that each degree of freedom represents one grid point. von Clarmann and Grabowski (2007) presented a method to constrain the retrieval to an appropriate grid by inspecting the averaging kernel of a prior constrained retrieval. As the magnitude of the diagonal values of the averaging kernel matrix are expressing the degrees of freedom of the retrieved profile at their certain heights, the way to find the appropriate retrieval grid is the following: The diagonal values of the averaging kernel matrix are summed up beginning from the first row until a natural number is reached, defining one box of the new coarse grid. In order to bound the entries of the state vector inside this box a first derivative operator is used increasing the strength such that all the values have to be the same. The solution is then given by

$$\hat{x} = (\mathbf{K}^T S_\epsilon^{-1} \cdot \mathbf{K} + \gamma^{-1} \cdot \mathbf{W})^{-1} \cdot (\mathbf{K}^T \cdot S_\epsilon^{-1} \cdot y) \qquad (4.41)$$

The choice of the grid

The choice of the grid for the inversion, in our case in space and time, is always a trade-off between avoidance of smoothing errors and the avoidance of asking for too much information from the measurements. While a coarse grid possibly misses certain structures of the retrieved state, a fine grid can produce unrealistic oscillations or desires a large contribution of the $a\ priori$ constraint to the retrieved state. The accessible resolution in altitude space, which is provided by the measurements can be learned from the averaging kernel matrix, given that the retrieval was performed on a sufficiently fine grid. In time space, the resolution is controlled by the initially defined grid and linear interpolation within this grid. The time spacing applied for all retrievals shown here is chosen to be 0.5 h. Concerning time

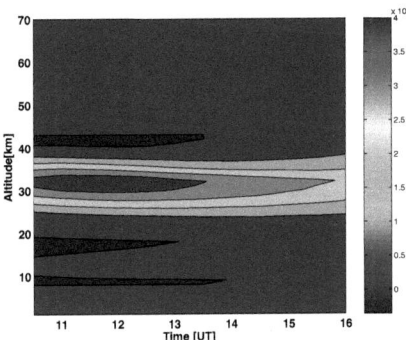

Figure 4.14: Gradient of the NO_2 concentration over time, as modeled with Labmos.

spacing, one has to consider that the grid is fine enough to illustrate the variation and that the change within one profile is not too large. Figure 4.14 shows the modeled gradient of NO_2. The change within the chosen time spacing $\Delta T_k = 0.5$h is <1% for all layers and times. Note, while above the subject was the change in concentration along the line-of-sight (space gradient), here the topic is the change of the concentration in time (time gradient, between two subsequent profiles in the retrieved state).

4.4.5 Characterization of the profile retrieval and error analysis

In this chapter a characterization of the retrieval by means of information content and resolution is given. We first estimate noise and retrieval errors analytically (Rodgers, 2000) and in the second part, through a sensitivity study, the forward model parameter error, caused by uncertainties of the viewing geometry. A first check of the model's ability to explain the measurements is given by the comparison of forward modeled and measured ΔSCDs, for which an example is shown in Figure 4.10. A quantitative measure for the consistency of measured and modeled ΔSCDs is the root mean square of the difference of both quantities.

$$RMS = \frac{\sqrt{(\sum_i y_i - y_{i,mod})^2}}{\overline{y_i}} \qquad (4.42)$$

where $y_{i,mod}$ is given by the product of $\mathbf{K_{1,j,k}}$ and $\vec{x_{j,k}}$. A more sophisticated characterization of the retrieval can be performed by calculating the averaging kernel matrix $\mathbf{A}$ with averaging kernels in its rows, which give the relation between the true value of state $\mathbf{x}$ and the retrieved state $\mathbf{\hat{x}}$.

$$\mathbf{\hat{x}} = \mathbf{x_a} + \mathbf{A}(\mathbf{x} - \mathbf{x_a}) + \epsilon. \qquad (4.43)$$

$\mathbf{A}$ is calculated from the weighting function and covariance matrices by

$$\mathbf{A} = \frac{d\mathbf{\hat{x}}}{d\mathbf{x}} = (\mathbf{K}^T \mathbf{S}_\epsilon^{-1} \mathbf{K} + \mathbf{S_a}^{-1})^{-1} \mathbf{K}^T \mathbf{S}_\epsilon^{-1} \mathbf{K}. \qquad (4.44)$$

In the ideal case $\mathbf{A}$ would be the identity matrix, meaning that the retrieved profile is resulting only from the measurements with an altitude and time resolution as good as the chosen grid, thus rendering any additional constraint such as the $a\ priori$ profile dispensable. In practice, the entries in the diagonal of the averaging kernels are often less than unity and expanded into neighboring layers, indicating contribution of the $a\ priori$ as well as a diminished resolution compared to the resolution of the retrieval grid.

The averaging kernels produced by this retrieval of time dependent trace gas profiles are not only a function of altitude but also of time. Therefore each element of the state vector has an averaging kernel that is two dimensional (Schofield et al., 2004). For the illustration of the time and altitude dependent contribution to one retrieved state vector element $x_{j,k}$, the averaging kernel is shown color coded over space and time. As an example Figure 4.15 shows the averaging kernel of the retrieval of the NO_2 concentration at 33 km, at 14 UT on June 30.

$$\mathbf{A} = \frac{d\mathbf{\hat{x}}_{33\,km,14\,UT}}{d\mathbf{x}_{j,k}} \qquad (4.45)$$

This example illustrates that the retrieval takes the information from a bounded region of space and time. In terms of resolution it is 1.5 km in altitude space and 1 hour in time. It should be pointed out, that altitude resolution is a result of measurement geometry, while time resolution is a result of the chosen time grid. The results from that retrieval are discussed in Section 5. As the time dependent contribution is somehow intuitive, for a better characterization of altitude resolution, we also show the averaging kernels of the retrieval of all altitudes at a certain time:

$$\mathbf{A} = \frac{d\mathbf{\hat{x}}_{j,14\,UT}}{d\mathbf{x}_{j,k}} \qquad (4.46)$$

This representation is displayed e.g. in Figures 5.44 and 5.46. The corresponding profiles are discussed in Section 5. The degrees of freedom describe the number of useful independent quantities that can be

CHAPTER 4. RETRIEVAL METHODS

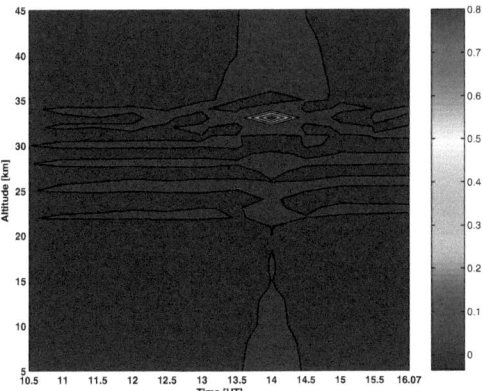

Figure 4.15: Two dimensional averaging kernel of the of the NO_2 concentration retrieval at 33 km, at 14 UT on June 30, 2005.

determined from a set of measurements. A measure of the degrees of freedom is the sum of the diagonal elements of the averaging kernel matrix. In order to quantify the independence of the *a priori* the sum of each averaging kernel (one line of $\mathbf{A}$) is derived. This is known as the area of an averaging kernel, or - for all averaging kernels - as the measurement response profile. When the area is approximately unity then most of the information on a spatial scale greater than the width of the averaging kernel is being supplied by the measurements rather than the a priori. In order to give a quantitative measure of the resolution of the profiles, the Backus-Gilbert (Backus, 1970) spread is calculated by

$$s_i = 12 \frac{\sum_j (i-j)^2 \mathbf{A}_{i,j}^2}{\sum_j \mathbf{A}_{i,j}^2} \tag{4.47}$$

Noise

The noise error of the retrieved profiles is a consequence of the noise in the measurements, which is quantified by the error of the DOAS analysis. Compared to that the Monte Carlo RTM noise is very small (around 1% in the considered SZA range) and is neglected henceforth. The retrieval noise covariance is given by

$$\hat{\mathbf{S}}_{\mathbf{noise}} = (\mathbf{K}^T \mathbf{S}_\epsilon^{-1} \mathbf{K})^{-1} \tag{4.48}$$

Smoothing

The retrieved profile can be regarded as smoothed by the averaging kernels. A particular smoothing error is commonly estimated by

$$\hat{\mathbf{S}}_{\mathbf{smooth}} = (\mathbf{A} - \mathbf{I}) \mathbf{S}_\mathbf{a} (\mathbf{A} - \mathbf{I})^T. \tag{4.49}$$

4.4. RETRIEVAL OF TRACE GAS PROFILES

but the true smoothing error can only be calculated by this formula if the true covariance $\mathbf{S_a}$ is known. Otherwise one has the questionable situation, that the error due to smoothing is lower, for lesser meaningful measurements. Consequently there is no smoothing error given in this work, but the retrieved profile has to be considered as smoothed by the averaging kernels.

Forward model Parameter error

Assuming that the forward model describes the physics of the measurements sufficiently precise, uncertainties remain for forward model parameters b_j. They are input parameters that influence the state but are not retrieved. For spectroscopic measurements in the atmosphere in general they include the composition of the atmosphere like aerosol load, cloud coverage and, most important in our case, viewing geometry.
The covariance due to each forward model parameter can be evaluated as

$$\mathbf{S_f} = \mathbf{G_y}^T \mathbf{K_b}^T \mathbf{S_b} \mathbf{K_b} \mathbf{G_y} \tag{4.50}$$

where $\mathbf{K_b}$ denotes the kernel concerning the particular forward model parameters b_j. The error arising due to aerosol load, derived by this method, is found to be negligible for our Limb scanning measurements. Unfortunately $\mathbf{K_b}$ for viewing geometry parameters is so far not an output of McArtim. In principle the kernel concerning the elevation angle K_α could be derived by numerical differentiation:

$$\frac{\frac{dy}{dx}x(\alpha) - \frac{dy}{dx}x(\alpha')}{\Delta \alpha} = \frac{dy}{d\alpha} \tag{4.51}$$

$$K_\alpha = \frac{dx}{dy}\frac{dy}{d\alpha} = \frac{dx}{d\alpha} \tag{4.52}$$

Since the dependency of the elevation angle is not linear, K_α needs to be derived iteratively. Alternatively, the corresponding error is characterized by performing sensitivity runs with respect to that quantity.

Sensitivity to unknown α^{**} oscillation

α^{**} oscillations (for the definition of α^{**} see section 3.2.2) the gondola are caused by wind shear acting on the balloon and gondola or by the azimuth stabilization system. Using a record of the attitude control parameters obtained during the MIPAS-B flight on June 14, a sensitivity study, on how the α^{**} oscillations propagates into the retrieved profile is performed. Different types of oscillations occurred during the flight, as can be seen in Figure 4.16. For the sensitivity study, the oscillations around 13:15 UT (type I) and 14:15 UT (type II) are treated separately, with two frequency regimes around 7 $\cdot 10^{-3}$ Hz and an amplitude of $\Delta \alpha^{**} = 1°$ and ≥ 1 Hz and an amplitude of $\Delta \alpha^{**} = 3°$, respectively. In order to quantify the error caused by these α^{**} oscillations, the line-of-sight $\mathbf{L_{i,j,osci}}$ for an oscillating gondola as well as the corresponding $y_{i,osci}$ (the product of $\mathbf{L_{i,j,osci}}$ and x_a) is calculated. We then retrieve an additional profile $\hat{x}_{osci}$ from $y_{i,osci}$, but the usual $\mathbf{L_{i,j}}$, which would be the result of a retrieval from measurements on an oscillating gondola and the RT assuming a stable gondola. x_a acts as the true profile in this study. The envelope of the differences among the two profiles (shown in Figure 4.16) builds the covariance $\mathbf{S_{osci}}$ caused by the unaccounted oscillations. The total error of the retrieved profile is then

80　　　　　　　　　　　　　　　　　　　　　　　　　　　　　CHAPTER 4. RETRIEVAL METHODS

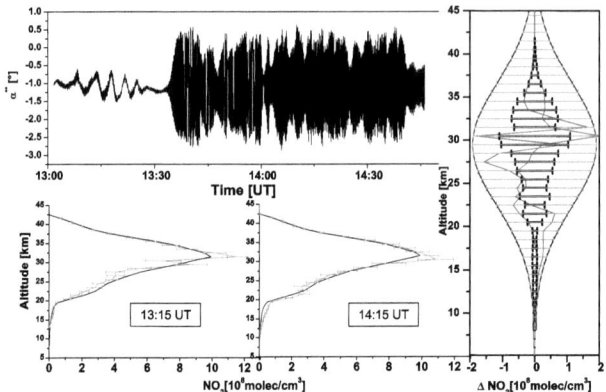

Figure 4.16: Upper left panel: Extreme cases (type I around 13:15 UT and type II around 14:15 UT) of α^{**} oscillation of the MIPAS-B gondola, as recorded by attitude control system. Lower left panel: Concentration of NO_2 retrieved from $\Delta SCDs$ expected on a calm flight (green) and on a gondola undergoing type I oscillation (black). Lower middle panel: Concentration of NO_2 retrieved from $\Delta SCDs$ expected on a calm flight (green) and on a gondola undergoing type II oscillation (black). Right panel: Differences of the retrieved profiles during both oscillation types shown on the left panel (red), envelope of the differences defined as an upper limit for $\mathbf{S_{osci}}$ (green error bars) and noise error $\mathbf{S_{noise}}$ (black error bars) for comparison.

given by the sum of the retrieval noise given in equation 4.48 and the oscillation error $\mathbf{S_{osci}}$.

$$\hat{\mathbf{S}} = (\mathbf{K}^T \mathbf{S}_\epsilon^{-1} \mathbf{K})^{-1} + \mathbf{S_{osci}} \qquad (4.53)$$

α^{**} oscillations with frequencies lower than the time resolution of the measurements cause oscillations in the retrieved profile, similar to a blurred photograph. They increase with increasing amplitude of initial oscillation and with the spatial distance of the observer to the observed object. α^{**} oscillations with frequencies higher than the time resolution of the measurements lead to a shift of the retrieved profile to higher layers because of the strong gradient in skylight radiance with tangent height. Thus the contribution of light coming from lower layers is large, compared to the contribution from higher layers.
This effect increases with the spatial distance between observer and object, due to the theorem on intersecting lines. As the float altitude of the considered balloon flight is around 35 km, both aspects are more serious for O_3 and BrO with maxima at around 26 km and 23 km, respectively than for NO_2 with a maximum at around 32.5 km, thus leading to a higher $\mathbf{S_{osci}}$ for O_3 ($\approx 15\%$) and BrO ($\approx 20\%$) than for NO_2 ($\approx 10\%$).
α^{**} oscillations are however typically much lower in amplitude than those used for this sensitivity study as can be concluded from an inspection of measured ΔSCDs (e.g. by comparison of Figures 5.5 and 5.40. The error arising from α oscillations can not be generalized, but have to be estimated carefully for

each situation. The error bars for the profiles presented in the following section are estimated considering 50% of S_{oscl} inferred from this study.

4.5 The retrieval of diurnal variation/chemical information

When retrieving profiles of diurnally varying trace gases one may think of several possible approaches. Firstly, to assume that the photochemistry is *a priori* known from chemical models and that a single profile varies in a predetermined way, e.g. as a scalar multiple or even in a more complex manner. Then, the photochemical variation can be treated as a forward model parameter (see 4.5). The second possible method, which is used here to account for the diurnal variation of the species, is to retrieve it. Sets of profiles defined on a time grid, which are retrieved describe the diurnal variation of the species. This approach does lead to an increased number of retrieval parameters compared to the latter method but avoids problems associated with assuming the diurnal variation as a forward model parameter and it allows direct comparison with chemical modelling. A third approach, shortly discussed here is to derive the parameters governing the diurnal variation, without going the way round through concentration profiles (see 4.5).
For example the diurnal variation of stratospheric NO_2 is governed by the photolysis of N_2O_5. Thus, it is possible to derive the photolysis frequency $J_{N_2O_5}$ from the gradient of its concentration (see 4.5).

Kalman Filter

Kalman Filtering is another method to constrain the inversion providing a connection between *a priori* profiles. Nevertheless the connection has to be known in advance (provided, that the time evolution of the state can be modeled (Rodgers, 2000). One can give more or less freedom to that by varying the covariance of the evolution error, but anyway the retrieved profile is influenced by the applied chemical model. Using Kalman filtering the temporal evolution is a forward model parameter, hence a quantity, which affects the measurements, but is not retrieved. It is an input and no output parameter. Kalman filtering is not an appropriate method to solve our inversion problem, because it gives constrains to the time evolution of the retrieved profile instead of absolute values, which is provided by the *a priori*. In conclusion the goal to retrieve the photolysis rate of N_2O_5, as the parameter determining the increase in NO_2, is not well supported by Kalman filtering.

$J_{N_2O_5}$ as a retrieval parameter

In principle a forward model for $J_{N_2O_5}$ can be set up and $K_{J_{N_2O_5}}$ can be derived by numerical differentiation:

$$\frac{\frac{dy}{dx}x(J_{N_2O_5}) - \frac{dy}{dx}x(J_{N_2O_5})}{\Delta J_{N_2O_5}} = \frac{dy}{dJ_{N_2O_5}} \tag{4.54}$$

$$K_{Ph} = \frac{dx}{dy}\frac{dy}{dJ_{N_2O_5}} = \frac{dx}{dJ_{N_2O_5}} \tag{4.55}$$

Forward modelling of $J_{N_2O_5}$

A similar approach to the above named is to retrieve the diurnal variation of NO_2 and to adjust the photochemical model to the observation. This method is applied in section 5.3.5.

Chapter 5

Results and Discussion

Within this thesis several measurement campaigns were performed (see table 3.2.3). The data shown in this chapter was obtained during the first deployments of the mini-DOAS instrument at a tropical site in northern Brazil (5° 4' S, 42° 52' W), in June 2005. The campaign took place at a private airport close to the city of Timon (Maranhao) near Teresina, the capital of Piaui. Timon is situated at the border between the federal states Piaui and Maranhao in Northeast Brazil and about 250 km south of the Atlantic coast (see figure 5.1).

Figure 5.1: Left panel: Satellite picture of north-east South America, with a mark at the measurement site. Adopted from: http://maps.google.de/maps. Right panel: Photography of the LPMA/DOAS payload during launch. Visible is the gondola hanging below the two auxiliary balloons, the much bigger main balloon is higher above.

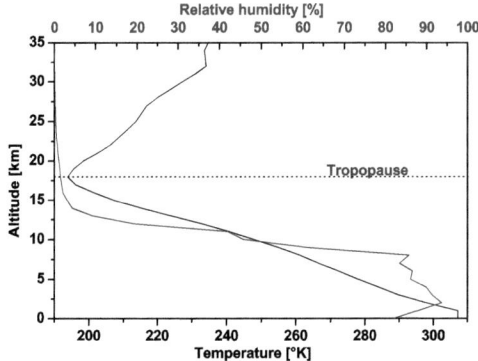

Figure 5.2: Relative humidity (blue) and temperature (black) as a function of altitude, measured by sonding at Teresina on June 17, 2005. The approximate height of the cold point tropopause is shown in purple dots ($\approx 17\,\text{km}$).

In the course of this campaign the mini-DOAS instrument was successfully brought to the stratosphere three times, and recovered in good order. Therefore it could be deployed subsequently on three different balloon payloads. Spectra from all three flights are analyzed for O_3, NO_2, BrO, O_4 and HONO. The results are shown in the following in chronological order. The relation between ΔSCD structure and viewing geometry is explained in more detail using as an example the first flight on the MIPAS-B payload.

Appropriate RTM requires meteorological data, because the air density and humidity are influencing the optical properties of the atmosphere and the ground. Temperature and humidity as measured by sonding during the campaign is shown in figure 5.2. Since the area around Teresina is savanna, the ground albedo is assumed to be 0.03 (Trishchenko et al., 2003). Aerosols are considered to be of organic type and parameters are set accordingly, while the extinction coefficient is derived from the measurements (see section 5.2.3).

In the vicinity of each flight the instrument was evacuated reaching a pressure of around 10^{-5} mbar and cooled to $0\,^\circ$C using a water-ice-mixture . Lamp spectra of a HgCd lamb and a halogen lamb were recorded in order to perform the wavelength calibration and to account for a changing sensitivity of the individual pixels (see chapter 4.1).

5.1 Observations from aboard MIPAS-B gondola on June 13, 2005

On June 13 the mini-DOAS instrument was integrated on board the MIPAS-B (Michelson Interferometer for Passive Atmospheric Sounding-Balloon, Oelhaf et al. (1991)) payload. This Fourier Transform Infra Red (FTIR) spectrometer allows limb emission sounding of vertical profiles of ozone and a considerable

5.1. OBSERVATIONS FROM ABOARD MIPAS-B GONDOLA ON JUNE 13, 2005

number of NO_x key radicals (NO, NO_2), NO_y reservoir species (HNO_3, N_2O_5, $ClONO_2$, and HO_2NO_2) as well as source gases (CH_4, N_2O, H_2O, CFC-11, CFC-12, CFC-22, CCl_4, CF_4, C_2H_6, and SF_6) simultaneously, at an altitude resolution of 2 to 3 km. Since the line-of-sight of the MIPAS-B spectrometer is kept at a constant SRAA of 90° our telescope was deployed parallel to the MIPAS-B telescope.

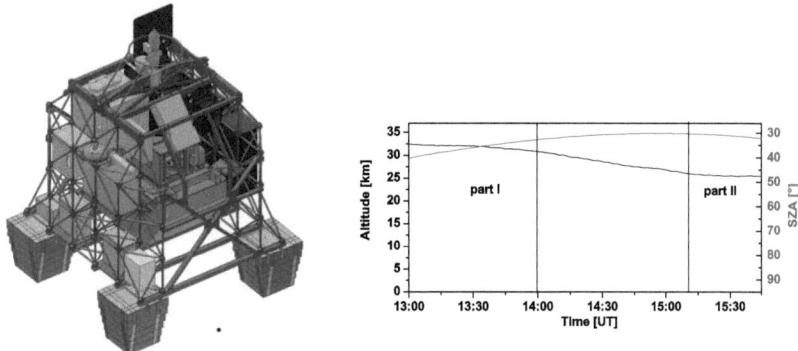

Figure 5.3: Left panel: MIPAS-B gondola with the position of the mini-DOAS instrument indicated by a green box [Sketch from Hans Nordmeyer, personal communication]. Right panel: Flight profile of the MIPAS-B gondola on June 13, 2005. Altitude (black) and SZA (red) versus time. Part I and part II denote periods of different float altitudes.

5.1.1 Flight conditions

The MIPAS-B payload was launched during night at 22:53 UT (19:53 LT) and floated until the next day at around 17:41 UT. This large balloon flight in the tropics could be performed utilizing the QBO wind system (see Figure 1.6).
Since scattered light measurements require sunlight, figure 5.3 shows the flight profile only from SZA around 90° on June 14. By then the MIPAS-B payload was floating at around 33 km height, starting a slow descent around 13 UT and reaching the new height of 25 km at around 15:30 UT. The mini-DOAS instrument was programmed to measure offset and dark current during night and perform limb scans during the day. Due to a software failure, the available data ranges only from 13 - 16 UT. During this time the instrument was performing limb scanning measurements, beginning at horizontal position and conducting 13 steps of 0.5° downwards. By comparing measured and modeled intensities the position of the telescope could be retrieved (see section 4.3), resulting in a sequence with starting elevation at $\alpha = -0.5°$ and ending at $\alpha = -6°$.
The radiative transfer of the observations is modeled using McArtim with forward model parameters like SZA (calculated) and SRAA and detector height obtained from the attitude system of the gondola. All measured quantities are affected by gondola α oscillations, mostly visible between 13:30 and 14:30 UT, when the balloon was governed to perform a slow descent and probably entering altitudes of varying

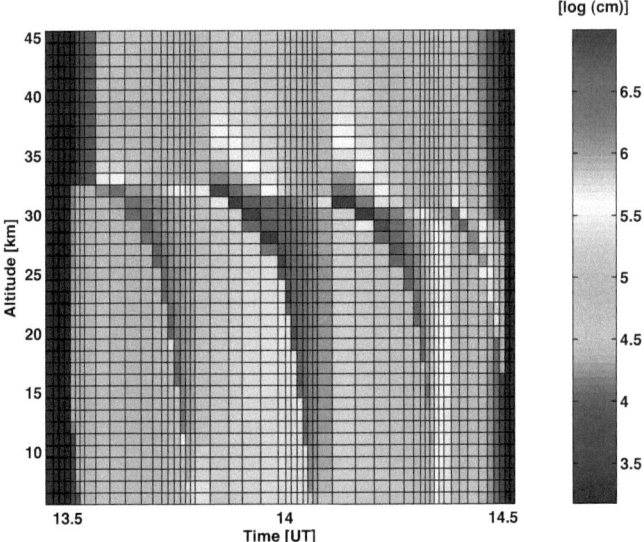

Figure 5.4: Logarithm of the weighting function $\mathbf{K}_{i,j}$ matrix elements for the retrieved NO_2 concentration profile at $t_k = 14$ [UT]. The example is for limb scanning measurements at around 33 km altitude with subsequent measurements in steps of 0.5° from α=-0.5° to α=-6° elevation angle from aboard the MIPAS-B payload, 2005.

wind regimes, resulting in serious oscillations caused by sheer winds. Since the MIPAS-B payload allows a record of several attitude control parameters including the elevation α^{**} (as defined in section 3.2.2, a sensitivity study is performed in order to quantify the error S_{osci} arising due to gondola α oscillations (see section 4.4.5). The conclusion is drawn, from a comparison of ΔSCDs retrieved from measurements during the MIPAS$_B$ flight and other balloon flights, that gondola oscillations during the MIPAS-B flight were unusually large and the derived error for the retrieved concentration is taken as an upper limit. The error bars for the profiles from measurements on the LPMA/IASI (e.g. Figure 5.50) and LPMA/DOAS (e.g. Figure 5.36) payload take into account 50% of S_{osci} derived from MIPAS-B α oscillations.

The time grid of the forward model is set up for a half-hourly retrieval of a profile throughout the measurement period. As an example, Figure 5.4 shows the contributions of the limb scanning measurements at around 33 km between t_i=13:30 UT and t_i=14:30 UT to the profile inferred for T_k=14 UT.

5.1.2 Measured ΔSCDs from MIPAS-B flight

In a first step measured and forward modeled ΔSCDs are compared for the targeted gases (Figure 5.5–5.9). The wavelike structure in the ΔSCDs occurring for all gases is a consequence of the scanning

5.1. OBSERVATIONS FROM ABOARD MIPAS-B GONDOLA ON JUNE 13, 2005

telescope. The shape of this structure is depending on the float height and the height of the maximum concentration of the particular gas.
In order to give a detailed explanation for the formation of this shape, the flight profile is divided into two parts of different float heights. The period from 13 UT to 14 UT is referred to as part I and the period from 15:30 to 16 UT referred to as part II. For each targeted gas the ΔSCDs are discussed for the two parts separately.

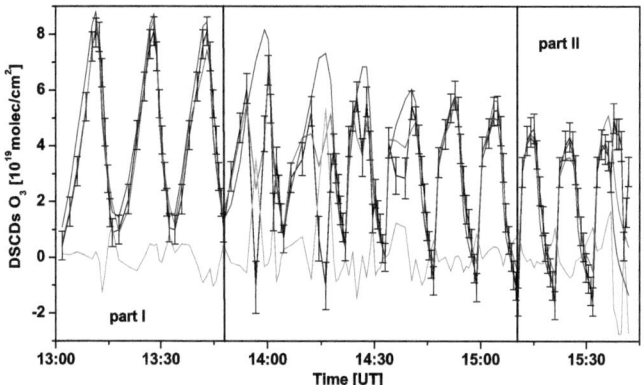

Figure 5.5: Measured (black), and simulated (red, blue) O_3 ΔSCDs from limb scanning measurements from aboard the MIPAS-B gondola on June 14, 2005. The forward modeled retrieved profiles are shown in red and the difference to the measured ΔSCDs (residual of the fit) is shown in green. Forward modeled *a priori* profiles (blue) are plotted for comparison.

ΔSCDs of O_3 from MIPAS-B flight

In the case of O_3 (Figure 5.5) the error of the spectral retrieval is small, compared to the amplitude of the observed ΔSCDs. The RMS of the residual is 3.1 and its structure mostly systematic, which can be explained by α oscillations or a remaining pointing error, both resulting in an error of the forward model parameter α (see section 4.4.5). Since the O_3 concentration maximum in the tropics is located at around 27 km, the ΔSCDs sequence starts in part I with a minimum, because the first elevation angle of $\alpha = -0.5°$ is leading to a light path (see Figure 5.4) above the ozone layer (see Figure 5.10) and therefore causing little absorption. With the viewing direction moving closer to the maximum of the O_3 layer the ΔSCDs increase, reach their maximum for an elevation of $\alpha = -3°$ and then decrease while scanning at lower elevation angles. The change in float altitude from 33 km in part I to 26 km in part II leads to a shift of the relation of elevation and ΔSCDs slope towards higher elevation. In part II the ΔSCDs at

$\alpha = -0.5°$ is already enhanced and the maximum is reached at $\alpha = -1.5°$. The reference spectrum for the spectral analysis of O_3 is taken at $\alpha = -6°$ in part II, where lowest ΔSCDs are measured due to a float below the O_3 concentration maximum combined with low elevation.

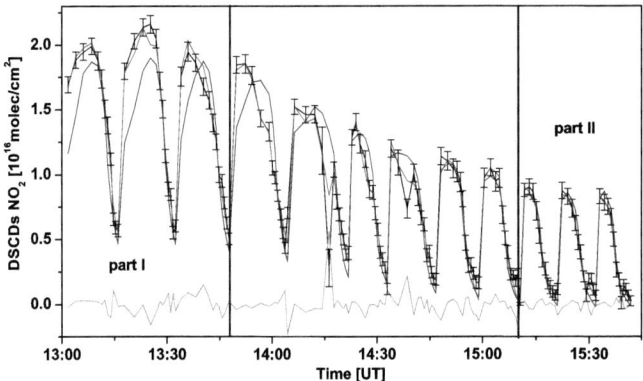

Figure 5.6: Measured (black), and simulated (red, blue) NO_2 ΔSCDs from limb scanning measurements from aboard the MIPAS-B gondola on June 14, 2005. The forward modeled retrieved profiles are shown in red and the difference to the measured ΔSCDs (residual of the fit) is shown in green. Forward modeled *a priori* profiles (blue) are plotted for comparison.

ΔSCDs of NO_2 from MIPAS-B flight
Measured and simulated NO_2 the ΔSCDs are shown in Figure 5.6. Here, the relative error is even lower than for O_3. Also the systematic structure of the residual is less because ΔSCDs of NO_2 with its maximum concentration at 33 km are less affected by α oscillations when the float altitude is also high (see Section 4.4.5). The concentration maximum at higher altitude as compared to O_3 leads to a shift of the relation of elevation and ΔSCDs slope towards higher elevation. In part I the ΔSCD at $\alpha = -0.5°$ is already enhanced and the maximum is reached at $\alpha = -2°$, which is similar to the ΔSCDs of O_3 in part I. This behaviour gives an indication of the ambiguity of the measurements if viewing geometry parameter are not well known, but also the possibility to retrieve one viewing geometry parameter if the others are known well (see Section 4.3).

5.1. OBSERVATIONS FROM ABOARD MIPAS-B GONDOLA ON JUNE 13, 2005

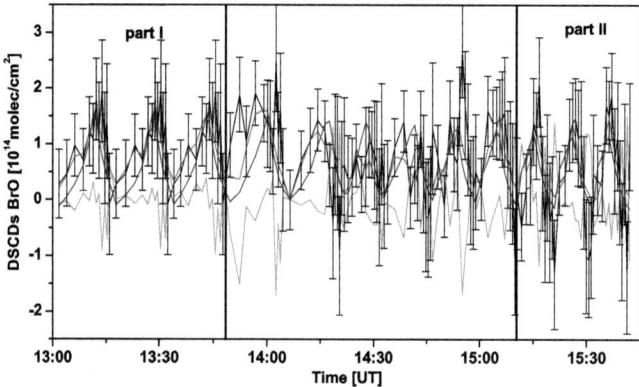

Figure 5.7: Measured (black), and simulated (red, blue) BrO ΔSCDs from limb scanning measurements from aboard the MIPAS-B gondola on June 14, 2005. The forward modeled retrieved profiles are shown in red and the difference to the measured ΔSCDs (residual of the fit) is shown in green. Forward modeled *a priori* profiles (blue) are plotted for comparison.

ΔSCDs of BrO from MIPAS-B flight

The results of the spectral analysis of BrO are shown in Figure 5.7. For BrO the ΔSCD error is high compared to its amplitude and also the systematic structure of the residual is higher compared to O_3 and NO_2 because ΔSCDs of BrO with its concentration maximum at 23 km are strongly affected by α oscillations when the float altitude is high (see Section 4.4.5). Also the amplitude of the ΔSCDs is low due to the same reason, that the instrument was far away from the layer to be observed.

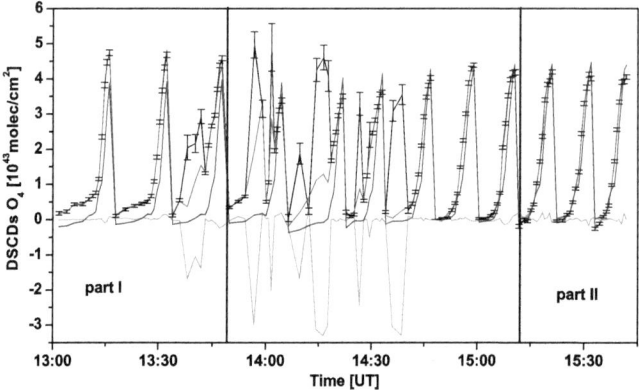

Figure 5.8: Measured (black), and simulated (red, blue) O_4 ΔSCDs from limb scanning measurements from aboard the MIPAS-B gondola on June 14, 2005. The forward modeled retrieved profiles are shown in red and the difference to the measured ΔSCDs (residual of the fit) is shown in green. Forward modeled *a priori* profiles (blue) are plotted for comparison.

ΔSCDs of O_4 from MIPAS-B flight

Figure 5.6 displays measured and modeled O_4 ΔSCDs. For low elevation angles the O_4 ΔSCDs are very high and the error is low compared to the amplitude, but the residual shows a high systematic structure, mainly between 13:30 and 14:40 UT, which is most probably caused by α oscillations of the gondola. Due to the shape of the O_4 profile (square exponential decline with height) the ΔSCDs are strongly affected by this α oscillations. The comparison of measured and modeled *a priori* O_4 ΔSCDs shows the validity of the forward model since the shape of the O_4 profile is known.

5.1. OBSERVATIONS FROM ABOARD MIPAS-B GONDOLA ON JUNE 13, 2005

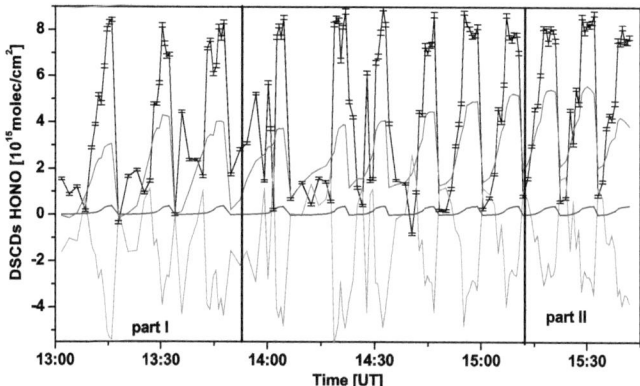

Figure 5.9: Measured (black), and simulated (red, blue) HONO ΔSCDs from limb scanning measurements from aboard the MIPAS-B gondola on June 14, 2005. The forward modeled retrieved profiles are shown in red and the difference to the measured ΔSCDs (residual of the fit) is shown in green. Forward modeled *a priori* profiles (blue) are plotted for comparison.

ΔSCDs of HONO from MIPAS-B flight

The results of the spectral retrieval of HONO ΔSCDs are shown in Figure 5.6. For HONO the residual is as high as the amplitude of the forward modeled ΔSCDs. The shape of the measured ΔSCDs indicates a certain shape for the concentration profile, but it is not sufficiently reproduced by the forward modeled ΔSCDs. This implies that the retrieval of a HONO concentration profiles shows large systematic errors.

5.1.3 Retrieved concentration profiles from MIPAS-B flight

In a second step the observations are inverted into trace gas concentration profiles, according to the method described in Section 4.4. The results of the inversion are shown in Figures 5.10 – 5.16. Each color plot is a composition of 7 height profiles. The upper panels on the left side show the retrieved trace gas concentration and the lower panels the area of the averaging kernels. The area of the averaging kernels gives an indication of the measurements contribution to the retrieved profiles (see section 4.4.5). The averaging kernels of the particular second limb profile (14 UT) are shown on the particular left side. Like the ΔSCDs, all retrieved concentration profiles are affected by the gondola α oscillations, which is most pronounced at around 14 UT. This effect is increasing with the spatial distance between observer and object, meaning the effect is strongest for O_4 and weakest for NO_2.

92   CHAPTER 5. RESULTS AND DISCUSSION

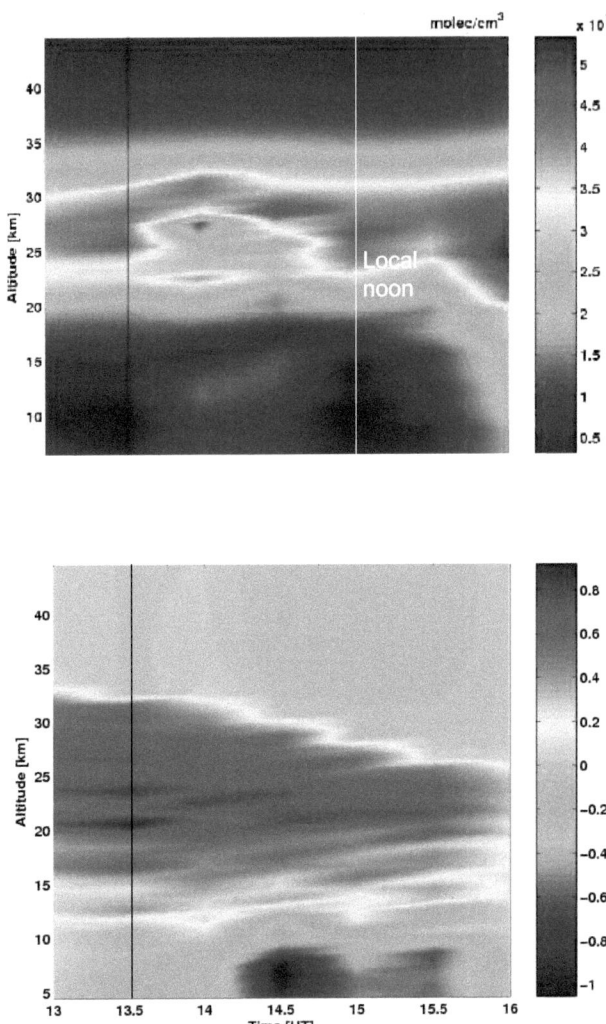

Figure 5.10: Retrieval of O_3 from balloon-borne measurements on board MIPAS-B payload on June 14, 2005. Upper panel: O_3 concentration versus altitude and time. Lower panel: Area of the averaging kernels versus altitude and time.

5.1. OBSERVATIONS FROM ABOARD MIPAS-B GONDOLA ON JUNE 13, 2005

Time series of O_3 concentration profiles from MIPAS-B flight

The upper panel of Figure 5.10 shows the retrieved O_3 concentration profiles versus time. The ozone concentration is highest in a layer between 21 and 30 km altitude, while a concentration maximum of around $5 \cdot 10^{12}$ molec/cm^3 is found at around 26 km altitude. The time variations are due to α oscillations of the gondola.

The area of the averaging kernels (lower panel) indicate that in the beginning the measurements yield the dominating contribution to the results between 20 and 30 km. The upper limit of the main contribution by the measurements is given by the float height and thus decreasing with time, as the gondola was performing a slow descent. The lower limit stays constant, due to the increasing importance of Rayleigh scattering in lower altitudes. Exemplarily, for the profile at 13:30 UT, which is designated in Figure 5.10 by a black line, the averaging kernels are shown Figure 5.11. They indicate a height resolution of around 1 km between 27 and 34 km altitude, decreasing slowly down to 5 km at 14 km altitude.

The O_3 retrieval from limb scanning measurements on June 14 from aboard MIPAS-B payload results in 48 degrees of freedom and a Shannon information content of 64 bits.

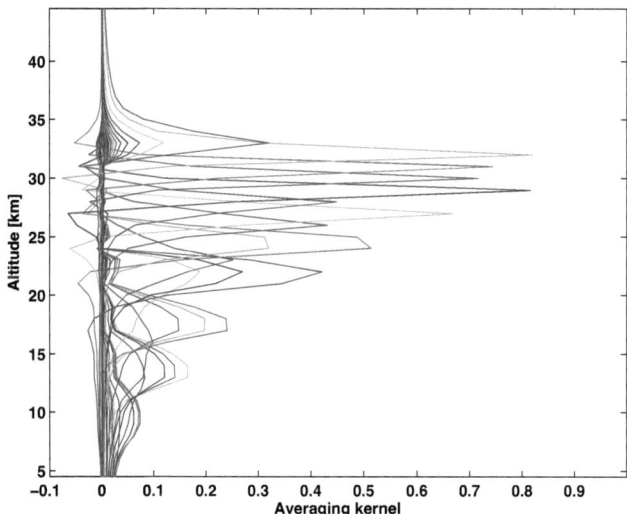

Figure 5.11: Averaging kernels of the second limb profile (indicated in Figure 5.10 by a black line)versus altitude.

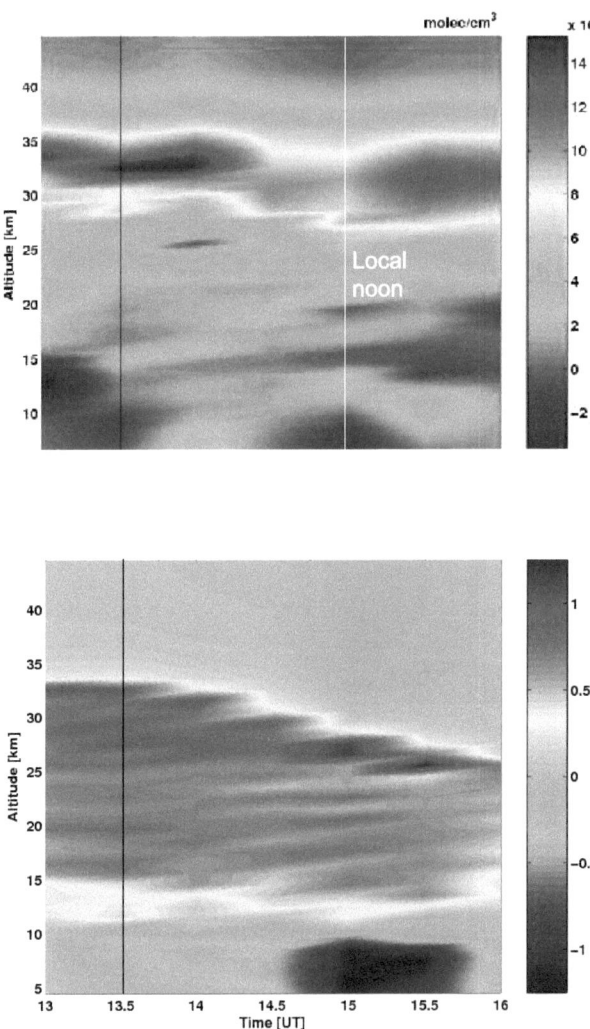

Figure 5.12: Retrieval of NO_2 from balloon-borne measurements on board MIPAS-B payload on June 14, 2005. Upper panel: NO_2 concentration versus altitude and time. Lower panel: Area of the averaging kernels versus altitude and time.

5.1. OBSERVATIONS FROM ABOARD MIPAS-B GONDOLA ON JUNE 13, 2005

Time series of NO_2 concentration profiles from MIPAS-B flight

The upper panel of Figure 5.12 shows the retrieved NO_2 concentration profiles versus time. The NO_2 concentration is highest in a layer between 28 and 36 km altitude, while a concentration maximum of around $1.5 \cdot 10^8$ molec/cm^3 is found at around 31 km altitude. The time variations are due to α oscillations of the gondola.

The measurements yield the dominant contribution to the retrieved profiles between around 15 and 34 km altitude, with the upper limit decreasing with decreasing float altitude. Exemplarily, for the profile at 13:30 UT, which is designated in Figure 5.12 by a black line, the averaging kernels are shown Figure 5.13. They indicate a height resolution of around 1 km between 27 and 34 km altitude, decreasing slowly down to 3 km at 14 km altitude.

The NO_2 retrieval from limb scanning measurements on June 14 from aboard MIPAS-B payload results in 70 degrees of freedom and a Shannon information content of 111 bits.

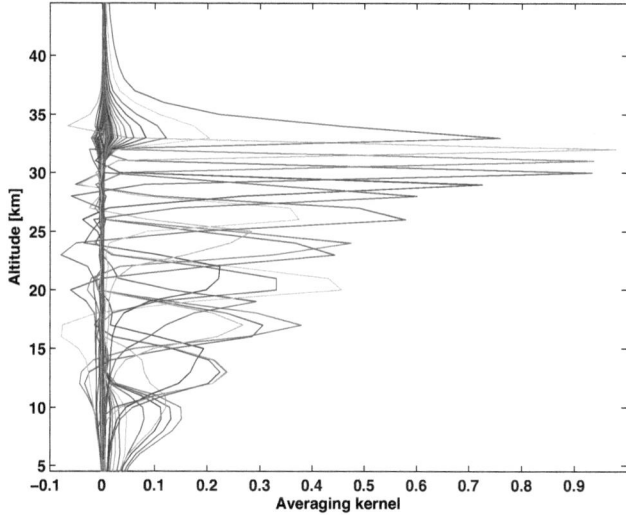

Figure 5.13: Averaging kernels of the second limb profile (indicated in Figure 5.12 by a black line) versus altitude.

96   CHAPTER 5. RESULTS AND DISCUSSION

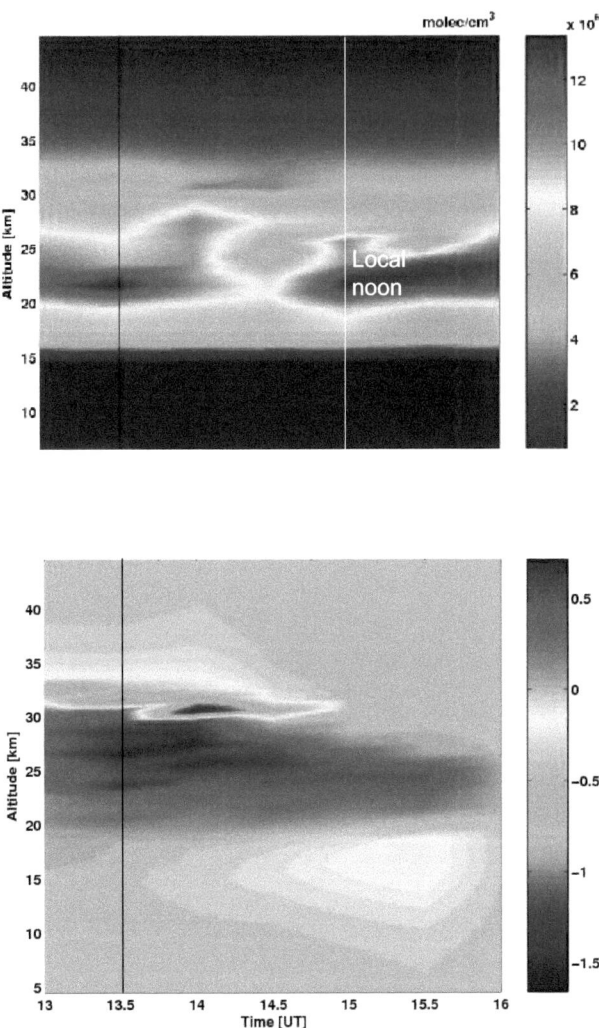

Figure 5.14: Retrieval of BrO from balloon-borne measurements on board MIPAS-B payload on June 14, 2005. Upper panel: BrO concentration versus altitude and time. Lower panel: Area of the averaging kernels versus altitude and time.

5.1. OBSERVATIONS FROM ABOARD MIPAS-B GONDOLA ON JUNE 13, 2005

Time series of BrO concentration profiles from MIPAS-B flight

The upper panel of Figure 5.14 shows the retrieved BrO concentration profiles versus time. The BrO concentration is highest in a layer between 20 and 27 km altitude, while a concentration maximum of around $13 \cdot 10^6$ molec/cm^3 is found at around 23 km altitude. The oscillations in the retrieved profile are due to α oscillations of the gondola.

The measurements yield the dominant contribution to the retrieved profiles between around 23 and 30 km altitude, with the upper limit decreasing with decreasing float altitude. Exemplarily, for the profile at 13:30 UT, which is designated in Figure 5.14 by a black line, the averaging kernels are shown Figure 5.15. They indicate a height resolution of around 4 km between 23 and 34 km altitude.

The BrO retrieval from limb scanning measurements on June 14 from aboard MIPAS-B payload results in 16 degrees of freedom and a shannon information content of 13.7 bits.

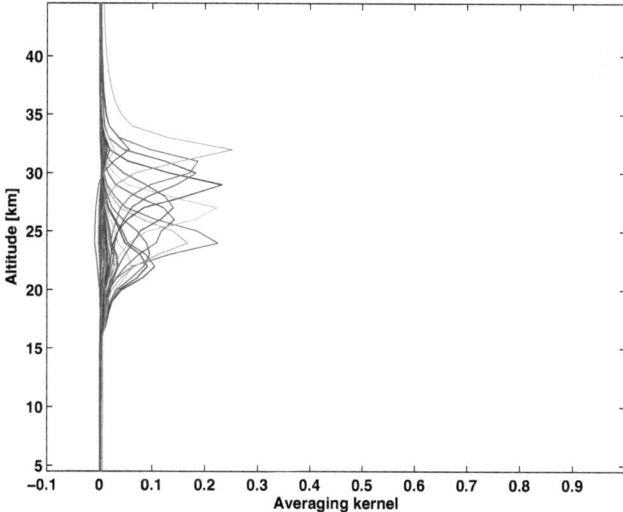

Figure 5.15: Averaging kernels of the second limb profile (indicated in Figure 5.14 by a black line) versus altitude.

98   CHAPTER 5. RESULTS AND DISCUSSION

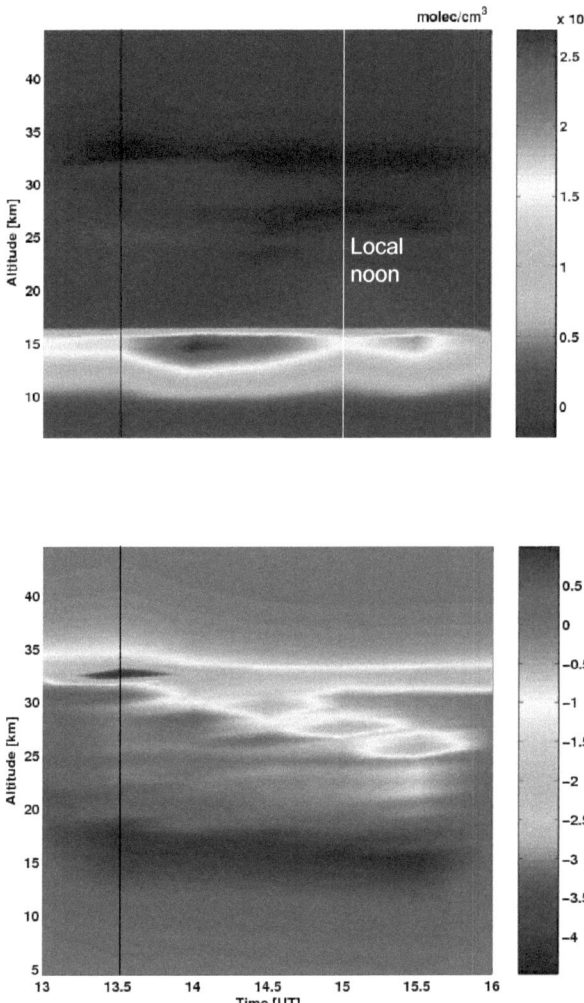

Figure 5.16: Retrieval of HONO from balloon-borne measurements on board MIPAS-B payload on June 14, 2005. Upper panel: HONO concentration versus altitude and time. Lower panel: Area of the averaging kernels versus altitude and time.

5.1. OBSERVATIONS FROM ABOARD MIPAS-B GONDOLA ON JUNE 13, 2005

Time series of HONO concentration profiles from MIPAS-B flight

The upper panel of Figure 5.16 shows the retrieved HONO concentration profiles versus time.
The retrieval of HONO concentration profiles is less reliable than for the other gases. As discussed above, the measurement/noise error S_ϵ, as well as the oscillation error S_{osci} is larger as compared to the other gases.
However, the resulting profiles show a maximum at around 15 km altitude throughout the measurement period. The temporal maximum at around 14 UT is coinciding with the largest α oscillations and therefore not realistic.
The area of the HONO retrieval is ranging between values of -4.5 to 1, with negative value indicating an unrealistic result in the retrieved profile. At altitudes between 10 and 20 km the area ranges between 0 and 1 and indicates a reliable result. Exemplarily, for the profile at 13:30 UT, which is designated in Figure 5.16 by a black line, the averaging kernels are shown Figure 5.17. They indicate a height resolution in the reliable range of of around 6 km.
The HONO retrieval from limb scanning measurements on June 14 from aboard MIPAS-B payload results in 29 degrees of freedom and a Shannon information content of 42 bits.
An attempt to interpret the abundance of HONO is abandoned in the following section.

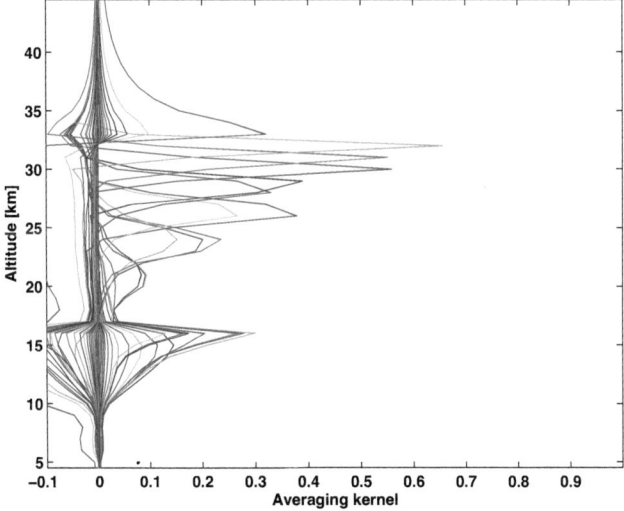

Figure 5.17: Averaging kernels of the second limb profile (indicated in Figure 5.16 by a black line) versus altitude.

5.1.4 Comparison of measured O_3 from MIPAS-B flight with in-situ ozone sonding

In order to validate the O_3 retrieval of concentration profiles, our ozone profile is compared to an in-situ measured O_3 profile. The O_3 sonde data are taken from an electrochemical cell launched from the ground station in the vicinity of the balloon flight. As tropical ozone is expected to be fairly constant with time, the time lag of one day between the two measurements appears to be acceptable. For the comparison, a mean profile of the above presented O_3 results (7 subsequent profiles) is used.

When comparing trace gas profiles at different altitude resolution the higher resolution profile needs to be degraded to the altitude resolution of the lower resolution profile. As the profile from the mini-DOAS instrument measurements is smoothed by its averaging kernels $\mathbf{A}$, the high resolution profile of the sonde is smoothed using the same averaging kernel matrix $\mathbf{A}$. The smoothed trace gas profile from the in-situ ozone sonde $\hat{x}_s$ is then given by

$$\hat{x}_s = x_a + \mathbf{A} \left(\hat{x}_h - x_a \right), \tag{5.1}$$

where x_a is the *a priori* profile used for the mini-DOAS retrieval (Connor et al., 1994; Hendrick et al., 2004b; Butz, 2006).

Figure 5.18 shows the comparison of the ozone profiles and the averaging kernels $\mathbf{A}$ of the mini-DOAS retrieval. The difference of both profiles is $\leq 10\%$ above 24km and on average 20% below this altitude. This good agreement of the two profiles gives confidence in the applied method of retrieving trace gas profiles from limb scanning measurements.

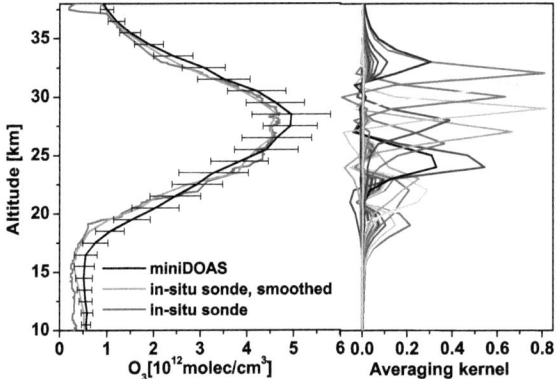

Figure 5.18: Left panel: Concentration profile of O_3, retrieved from balloon-borne measurements on board MIPAS-B payload on June 14, 2005 (black) and from in-situ sonde in the vicinity of the balloon flight (red), smoothed with the averaging kernels of the mini-DOAS retrieval(green). Right panel: Averaging kernels of the mini-DOAS retrieval of O_3.

5.1.5 Detection of lightning NO_x and HONO during the MIPAS-B flight

Unexpectedly, the spectral retrieval of the spectra from MIPAS-B flight give clear indication of HONO absorption. Albeit the residual is structured and exceeding the differential structure of HONO only little, a comparison spectral fits performed in different wavelength intervals and the inclusion of absorption cross sections from different gases leads to a stable result of HONO absorption in the measured spectra. Spectra taken during the other balloon flights of the campaign do not show a spectral signal of HONO. The distribution of the ΔSCDs over the elevation indicates the appearance of HONO in the upper troposphere.

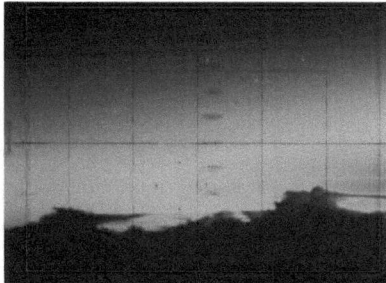

Figure 5.19: High clouds as seen from the Mipas-B star camera looking to NNE at about 40 min before sunrise (9:15 UT) on June 14 2005. Overshooting cloud tops are at about 12 km, a thin cirrus cloud is at about 14 km. Higher up in the clear stratosphere some stars are still visible.

Figure 5.20 shows a map of north-eastern Brazil with the viewing direction of the mini-DOAS instrument on the left panel and data from the World-Wide Lightning Location Network (WWLLN) on the right panel. Here, the green star shows the location of the measurement. The squares indicate lightning discharges with the colour coding universal time. Since WWLLN has very low sensitivity ($\approx 1\%$) for this region (Rodger, 2006), the reported detection of lightning indicates very strong events. Lightning is detected north, west and south of the location of the measurement. Lightning detection and the viewing direction of the measurements match for the last hour of the measurement period, but it is also probable that there was lightning throughout the entire measurement period. Since no lightning is reported for the other balloon flights within that campaign, one may speculate as to whether lightning may be the cause for HONO detection. This assumption is encouraged by the observation of a tropospheric NO_2 maximum of about $7 \cdot 10^8$ molec/cm^3 in around 13 km altitude (see Figure 5.12), with the inconsistency, that this enhanced NO_2 is not only pronounced around 15:30 UT. NO_x is produced by lightning, as described in section 1.5.1 (Pickering, 1998). The reactions forming HONO are listed in section 1.6. Since the reaction of exited NO_2 (marked with a star) and water is negligible, under upper tropospheric conditions and the observation of O_4 (see Figure 5.8) gives no indication for aerosols or thick clouds in the light path,

102   CHAPTER 5. RESULTS AND DISCUSSION

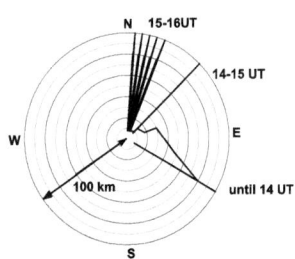 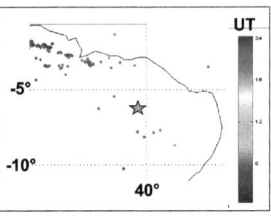

Figure 5.20: Left panel: Azimuthal viewing direction (black lines) of the mini-DOAS instrument during MIPAS-B flight (times given in UT). Right panel: Map of north-eastern Brazil with real-time locations of lightning discharges from the World-Wide Lightning Location Network (WWLLN) for the same day. The colour codes time in UT. The green star shows the location of the measurements.

we analyse the observed amount of HONO on the assumption of photo stationary conditions.
Through the link from NO_2 to NO all gases belonging to the photo stationary state are measured by the mini-DOAS instrument despite of OH. Hence our observation results in a constraint on the OH abundance in the related altitude range. Reaction constants k_{NO+O_3} and k_{NO+OH} are calculated using the measured temperature and pressure (Figure 5.2) and JPL 2006 values. According to reactions R1.4 and R1.5 and R1.35 to R1.37 and assuming a diurnal mean concentration of OH as given in (Salzmann, 2008) the expected HONO abundance is calculated by

$$[HONO] = \frac{k_{NO+OH} \cdot J_{NO_2}[NO_2] \cdot [OH]}{J_{HONO} \cdot k_{NO+O_3} \cdot [O_3]} \tag{5.2}$$

This concentration profile (calculated for all relevant altitudes) of HONO is applied as *a priori* profile for the inversion (see section 4.4.3) and is compared to the measured concentration of HONO in Figure 5.21 on the left panel.
The other way round the OH abundance, which is necessary to explain our HONO observation is calculated by

$$[OH] = \frac{J_{HONO}[HONO] \cdot k_{NO+O_3} \cdot [O_3]}{k_{NO+OH} \cdot J_{NO_2} \cdot NO_2} \tag{5.3}$$

and compared to the expected OH concentration as given in Salzmann (2008); Pickering (1998) (figure 5.21, right panel). There are now two possibilities to explain the mini-DOAS measurements.
Either there were additional sources of OH compared to the expected amounts by Salzmann (2008), or there were additional reactions forming HONO compared to the considered reactions. A candidate for the former explanation is the oxidation of acetone or other VOCs as emission from living plants or decaying litter in the rain forest. They can act as an OH source in the upper troposphere after being transported from the amazonian planetary boundary layer by deep convection (Poeschl, 2001). Heterogeneous reactions of NO_2 and H_2O might be an additional source of HONO. Here clouds or aerosols could

5.1. OBSERVATIONS FROM ABOARD MIPAS-B GONDOLA ON JUNE 13, 2005

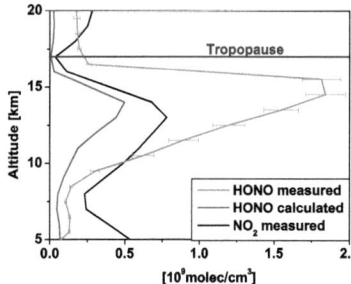
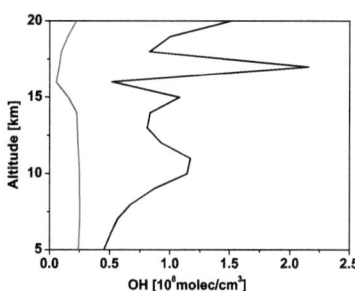

Figure 5.21: Left panel: Observed NO_2 (black), calculated HONO, which is used as *a priori* for the inversion (red) and retrieved HONO (green), height of the tropopause (blue). Right panel: Concentration of OH (black, calculated by equation 5.3) which is necessary to explain the measured amount of HONO and expected concentration of OH (red) (Salzmann, 2008).

provide surfaces, even though the O_4 observation do not indicate such conditions. HONO has been measured in thunderstorms also by Naumann and Dix et al. (2009) (70 ppb) at around 10 km altitude, with concentration about half as much as reported here.

5.2 Observations from aboard LPMA/DOAS gondola on June 17, 2005

On June 17 the mini-DOAS instrument was integrated on board the LPMA/DOAS (Laboratoire de Physique Moléculaire et Applications and Differential Optical Absorption Spectroscopy) gondola. The LPMA/DOAS balloon gondola is based on a gondola developed for astronomical observations by the Observatoire de Genève and was further optimized for atmospheric measurements by Camy-Peyret et al. (1995). The gondola can be stabilized in azimuthal direction by an alignment to the magnetic field of the earth with a gyroscope and rotated with respect to the much larger balloon. The fine-pointing is performed by a sun tracker (Hawat et al., 1995), which provides the infrared Fourier Transform Interferometer and the DOAS UV/vis spectrograph with a parallel solar beam. The infrared Fourier Transform Interferometer (FTIR) operated by the french LPMA team is a Michelson interferometer with plane mirrors. The instrument can detect several gases like $ClONO_2$, HNO_3, O_3, CH_4, N_2O, NO, and H_2O, which are covered by the HgCdTe detector (mid-IR) and HCl, NO_2, CH_4, and HF by the InSb detector (near-IR) (Camy-Peyret et al., 1995; Payan et al., 1998, 1999).

The parallel light beam of the sun tracker is also used by a direct sunlight DOAS instrument, operated by our group. This spectrometer is optimized for airborne applications and was designed and developed by Ferlemann (1998); Ferlemann et al. (2000) and Harder et al. (1998). The instrument provides faintly good stability of the spectral imaging and insignificant thermal drift of the spectroscopic system. The main difference compared to the mini-DOAS instrument lies in the strict and well defined radiative transfer (see section 3.2.2).

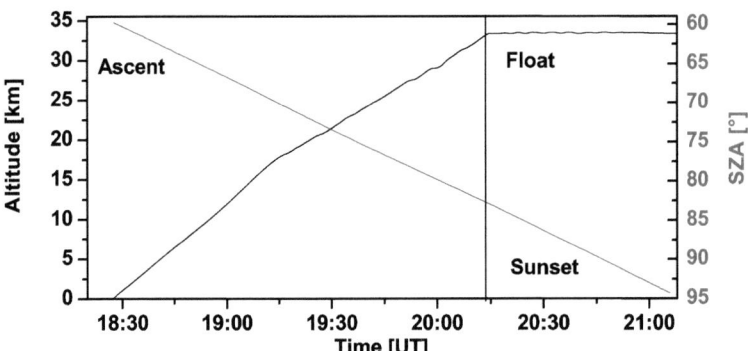

Figure 5.22: Flight profile of the LPMA/DOAS gondola on June 17, 2005. Altitude (black) and SZA (red) versus Universal Time.

5.2. OBSERVATIONS FROM ABOARD LPMA/DOAS GONDOLA ON JUNE 17, 2005

5.2.1 Flight conditions

The LPMA/DOAS payload was launched on June 17 shortly before 18:30 UT, reaching float altitude of 33 km at 20:15 UT and was the cut was performed after sunset, which occurred around 20:45 UT at float height. Throughout the deployment the mini-DOAS instrument measured successfully.
The telescope was oriented to the horizon at fixed elevation angle α during balloon ascent. When the balloon reached float altitude, the telescope was commanded to automatically scan 12 different elevations in steps of $0.5°$. Afterwards the elevation is retrieved to $\alpha = -2°$ during ascent and a sequence from $\alpha = -1°$ to $\alpha = -6.5°$ during float. For ascent observations this allows the highest sensitivity to balloon altitude or slightly below (see Figure 5.23) and during float a sensitivity to altitudes between 40 and 5 km.

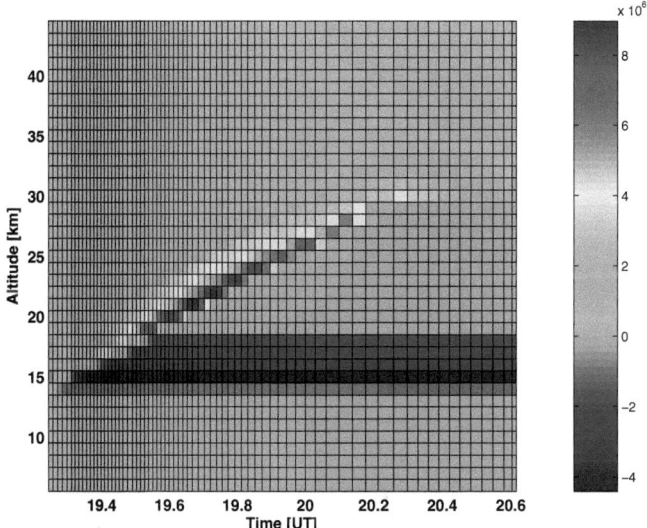

Figure 5.23: Logarithm of the weighting function $\mathbf{K}_{i,j}$ matrix elements for the retrieved NO_2 concentration profile at $t_k = 19:42$ [UT]. The example shows measurements during balloon ascent from 15 to 31 km altitude with a telescope elevation of $\alpha = -2°$ from aboard the LPMA/DOAS payload.

Modeling of the radiative transfer at high SZA and low elevation can result in unrealistic values due to numerical problems. Some measurements from this flight are excluded because their respective BoxAMFs could not be calculated or yield infinity values. This is the case for $\alpha = -6°$ and $\alpha = -6.5°$ at SZA = $92°$ and for $\alpha = -5.5°$, $\alpha = -6°$ and $\alpha = -6.5°$ at SZA = $94°$.

5.2.2 Measured ΔSCDs from LPMA/DOAS flight

In a first step measured and forward modeled ΔSCDs are compared for the targeted gases (Figure 5.24–5.26). For trace gases having their maximum in the stratosphere the ΔSCDs are low during take of and increase at altitudes when the balloon reaches the particular layer. occurring for all gases is a consequence of the scanning telescope.

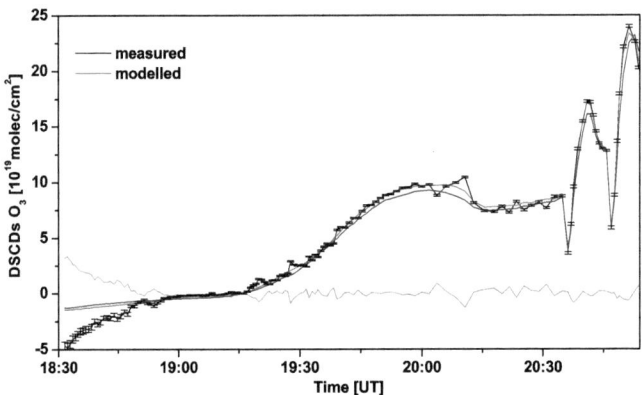

Figure 5.24: Measured (black), and simulated (red, blue) O_3 ΔSCDs from Limb scanning measurements from aboard the LPMA/DOAS gondola on June 17, 2005. The forward modeled retrieved profiles are shown in red and the difference to the measured ones (residual of the fit) is shown in green. Forward modeled *a priori* profiles (blue) are plotted for comparison.

ΔSCDs of O_3 from LPMA/DOAS flight

Figure 5.24 shows a comparison of measured and forward modeled O_3 ΔSCDs. The agreement is very good, except for the very beginning, when the balloon was just taking of. This disagreement of measured and modeled ΔSCDs near the ground can be due to unaccounted aerosols in the boundary layer. Around 19:20 UT the O_3 ΔSCDs begin to increase, when the balloon reaches the tropopause region. The first maximum is reached around 20 UT, at a balloon height of around 27 km, where the maximum concentration of ozone is expected. Then ΔSCDs decrease again until 20:15 UT at a balloon height of 31 km. The peak at around 20:10 UT is probably due to an oscillation of the gondola, as it is visible in the flight profile (Figure5.22) and also in the ΔSCDs of BrO. The following two peaks are due to Limb scanning, which started at 20:40 UT.

5.2. OBSERVATIONS FROM ABOARD LPMA/DOAS GONDOLA ON JUNE 17, 2005

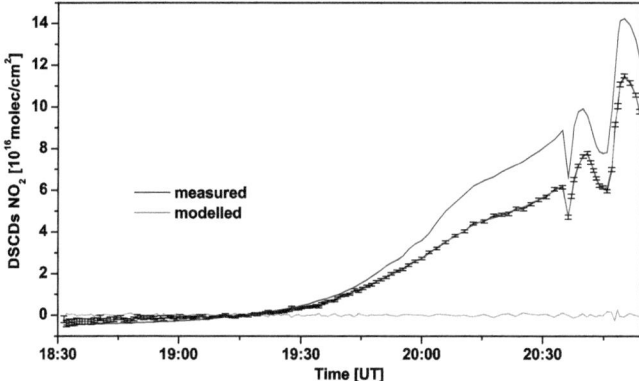

Figure 5.25: Measured (black), and simulated (red, blue) NO_2 ΔSCDs from Limb scanning measurements from aboard the LPMA/DOAS gondola on June 17, 2005. The forward modeled retrieved profiles are shown in red and the difference to the measured ΔSCDs (residual of the fit) is shown in green. Forward modeled a priori profiles (blue) are plotted for comparison.

ΔSCDs of NO_2 from LPMA/DOAS flight

Figure 5.25 shows measured and forward modeled NO_2 ΔSCDs. The agreement is very good and the residual is small for the whole measurement period. NO_2 ΔSCDs begin to increase around 19:20 UT which corresponds to a tangent height of 18 km. The following two peaks are due to Limb scanning, which started at 20:40 UT.

108   CHAPTER 5. RESULTS AND DISCUSSION

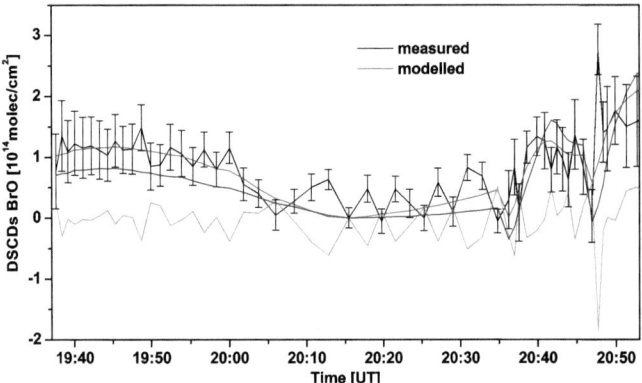

Figure 5.26: Measured (black), and simulated (red, blue) BrO ΔSCDs from Limb scanning measurements from aboard the LPMA/DOAS gondola on June 17, 2005. The forward modeled retrieved profiles are shown in red and the difference to the measured ΔSCDs (residual of the fit) is shown in green. Forward modeled *a priori* profiles (blue) are plotted for comparison.

ΔSCDs of BrO from LPMA/DOAS flight

Figure 5.26 shows measured and forward modeled BrO ΔSCDs. The BrO measurements are only considered in the time frame from 19:36 UT to 20:54 UT since the spectral retrieval here is most reliable. Anyway, the error corresponding to the spectral analysis is very high and is exceeding the absolute value most of the time (below detection limit). BrO ΔSCDs are exceeding detection limit from around 19:40 UT, corresponding to a tangent height of 20 km. Their further developing also looks reasonable.

5.2. OBSERVATIONS FROM ABOARD LPMA/DOAS GONDOLA ON JUNE 17, 2005

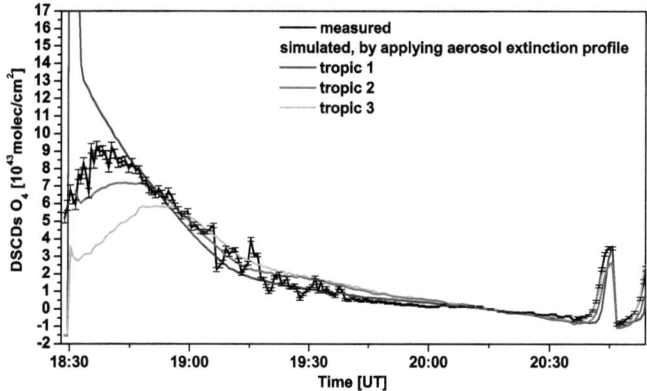

Figure 5.27: Measured (black), and simulated (blue, pink, cyan) O_4 ΔSCDs from Limb scanning measurements from aboard the LPMA/DOAS gondola on June 17, 2005. The different simulated ΔSCDs are obtained by applying different aerosol extinction profiles (figure 5.28) for the RTM.

ΔSCDs of O_4 from LPMA/DOAS flight

Figure 5.27 shows measured and forward modeled O_4 ΔSCDs. Due to the shape of the O_4 concentration profile the ΔSCDs are high at low balloon altitudes and decreasing during balloon ascent. On the basis of the comparison of modeled and measured O_4 ΔSCDs a tropospheric aerosol extinction profile is extracted as it is explained in the following section.

5.2.3 Aerosol extinction profile

Atmospheric stray light measurements during balloon ascent are sensitive to tropospheric aerosols and the occurrence of clouds. Aerosols are a potential difficulty in modeling the atmospheric radiative transfer as their properties are not precisely known. For the stratospheric aerosol extinctions satellite measurements exist which can be used for the RT modeling. For tropospheric aerosols, however, no measurements are available for the time of our balloon flights.

The parameterization of aerosols in the radiative transfer as implemented in McArtim is based on three parameters: the single scattering albedo, the Henyey-Greenstein parameter g for the scattering phase function and the extinction coefficient. For the stratosphere an aerosol extinction profile as indicated by direct sun measurements on the same flight (Butz, 2006) is applied. The parameter for the phase function g is set to 0.85 and the single scattering albedo to 0.99, according to a mixture of organic aerosol and smoke, which can be expected in the area of the measurements. Since tropospheric aerosols are highly variable, several aerosol extinction profiles as shown in Figure 5.28 are tested.

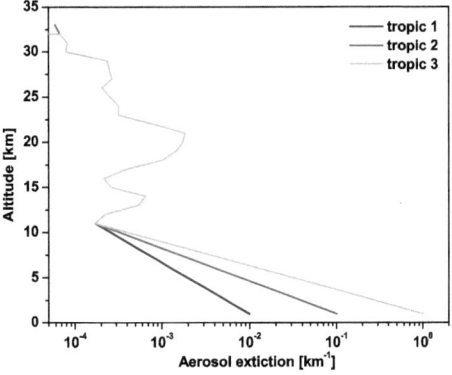

Figure 5.28: Aerosol extinction profiles as applied for the RTM of Limb scanning measurements from aboard the LPMA/DOAS gondola, for 490nm.

No inversion of the forward model is performed (Friess et al., 2006), but the aerosol extinction profile is iteratively changed until measured and modeled O_4 ΔSCDs fit together.

It turns out, that the aerosol extinction profile tropic 2 results in simulated O_4 ΔSCDs that fit best to the measured ones. In conclusion this profile is used for radiative transfer modeling of all measurements discussed in this thesis, where it is to note, that Limb scanning in around 33 km altitude is not sensitive to tropospheric aerosols.

Another approach to retrieve information on scattering in the atmosphere is to compare modeled and measured intensities. Figure 5.29 shows in the left panel the logarithm of the ratio of I_i and I_{ref}, which

5.2. OBSERVATIONS FROM ABOARD LPMA/DOAS GONDOLA ON JUNE 17, 2005

is also sensitive to aerosols (see section 4.1). The left shows the retrieved aerosol profile in different iterations. The last iteration (blue line) is in good agreement to the extinction profile extracted from the O_4 measurements.

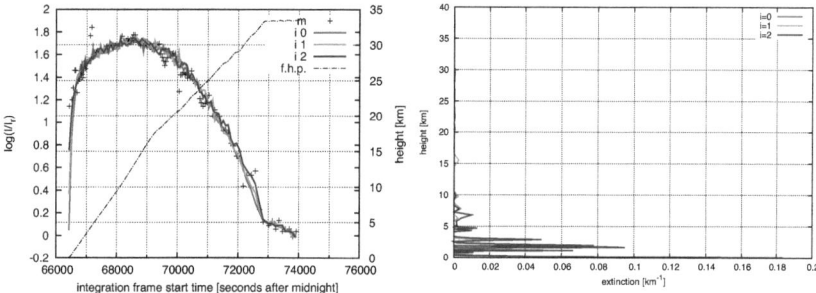

Figure 5.29: Left panel: Logarithm of the ratio of measured intensity to reference intensity compared to the ratio of modeled intensity to reference intensity. Right panel: Retrieved aerosol extinction profile. [Plots from Tim Deutschmann, personal communication]

5.2.4 Retrieved concentration profiles from LPMA/DOAS flight

In a second step the observations are inverted into trace gas concentration profiles, according to the method described in Section 4.4. The results of the inversion are shown in Figures 5.30 – 5.34. Each color plot is a composition of 4 height profiles. The upper panels on the left side show the retrieved trace gas concentration and the lower panels the area of the averaging kernels. The area of the averaging kernels gives an indication of the measurements contribution to the retrieved profiles (see section 4.4.5). The averaging kernels of the particular second limb profile (19:46 UT) are shown on the particular left side.

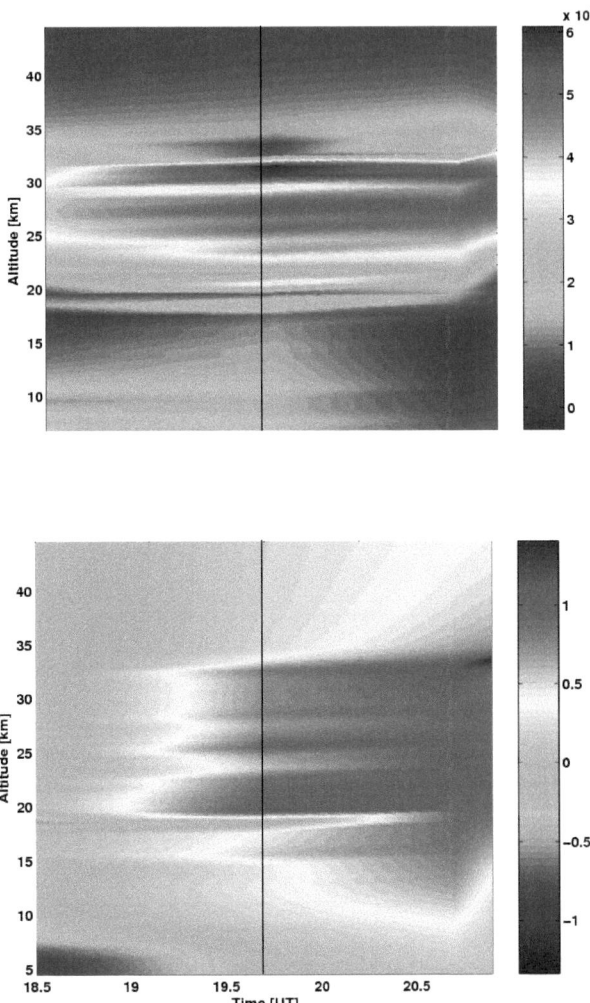

Figure 5.30: Retrieval of O_3 from balloon-borne measurements on board LPMA/DOAS payload on June 17, 2005. Upper panel: O_3 concentration versus altitude and time. Lower panel: Area of the averaging kernels versus altitude and time.

5.2. OBSERVATIONS FROM ABOARD LPMA/DOAS GONDOLA ON JUNE 17, 2005

Time series of O_3 concentration profiles from LPMA/DOAS flight

The upper panel of Figure 5.30 shows the retrieved O_3 concentration profiles versus time. The ozone concentration is highest in a layer between 23 and 32 km altitude, while a concentration maximum of around $5.5 \cdot 10^{12}$ molec/cm^3 is found at around 27 km altitude.

The area of the averaging kernels (lower panel) shows the contribution of the measurements to the retrieved profiles (see section 4.4.5) for the subsequent profiles. The contribution of the measurements is increasing towards sunset as light paths are becoming longer for lower SZA. Exemplarily, for the profile at 19:46 UT, which is designated in Figure 5.30 by a black line the averaging kernels are shown Figure 5.31. Since this profile is retrieved from balloon ascent measurements, the height resolution exhibits values of around 1 km between 18 and 31 km altitude and around 4 km at 5 km altitude. The retrieval of ozone from ascent and Limb scanning measurements from ascent and Limb scanning measurements from aboard LPMA/DOAS payload results in 33 degrees of freedom and a Shannon information content of 48 bits.

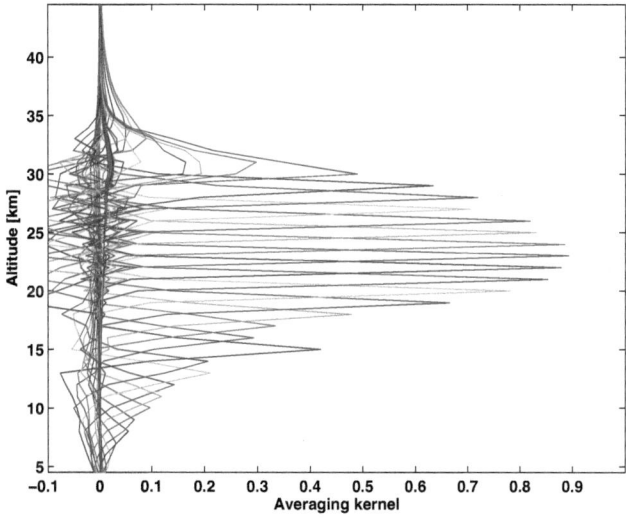

Figure 5.31: Averaging kernels of the second limb profile (indicated in Figure 5.30 by a black line) versus altitude.

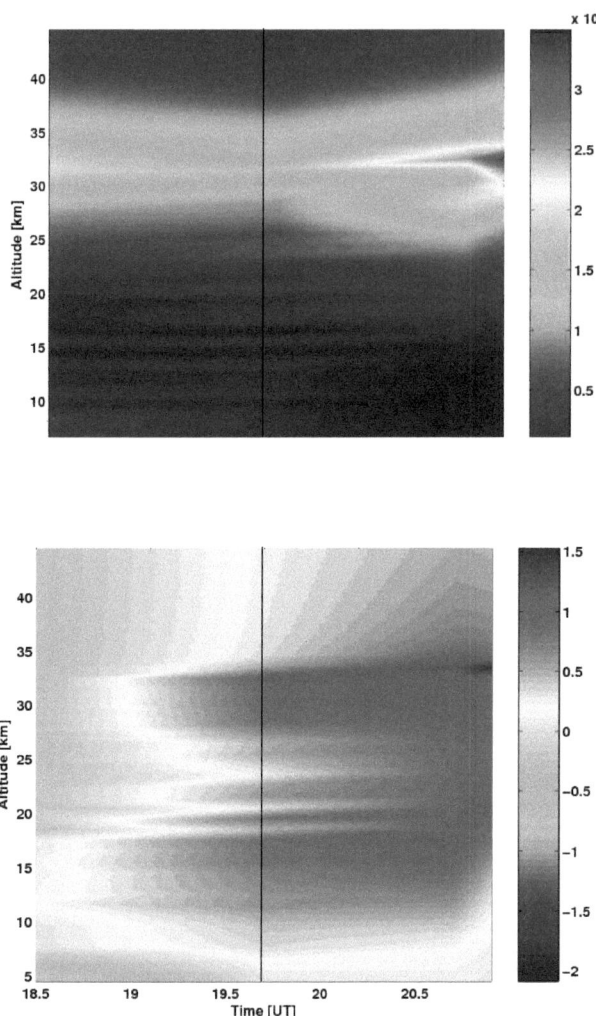

Figure 5.32: Retrieval of NO_2 from balloon-borne measurements on board LPMA/DOAS payload on June 17, 2005. Upper panel: NO_2 concentration versus altitude and time. Lower panel: Area of the averaging kernels versus altitude and time.

5.2. OBSERVATIONS FROM ABOARD LPMA/DOAS GONDOLA ON JUNE 17, 2005

Time series of NO_2 concentration profiles from LPMA/DOAS flight

The upper panel of Figure 5.32 shows the retrieved NO_2 concentration profiles versus time. The NO_2 concentration is highest in a layer between 28 and 36 km altitude. A strong gradient is observed in the profiles from $SZA = 60°$ to $SZA = 90°$ (for SZA see Figure 5.22). While at 19:46 UT the concentration maximum exhibits around $2 \cdot 10^9$ molec/cm^3 it reaches around $3.5 \cdot 10^9$ molec/cm^3 at 20:46 UT, when also a second maximum has evolved at around 27 km altitude.

The contribution of the measurements is increasing towards sunset as light paths are becoming longer for lower SZA. Exemplarily, for the profile at 19:46 UT, which is designated in Figure 5.32 by a black line the averaging kernels are shown Figure 5.33. Since this profile is retrieved from balloon ascent measurements, the height resolution exhibits values of around 1 km between 18 and 31 km altitude and around 4 km at 5 km altitude. The retrieval of NO_2 from ascent and Limb scanning measurements from aboard LPMA payload results in 37 degrees of freedom and a Shannon information content is 56 bits.

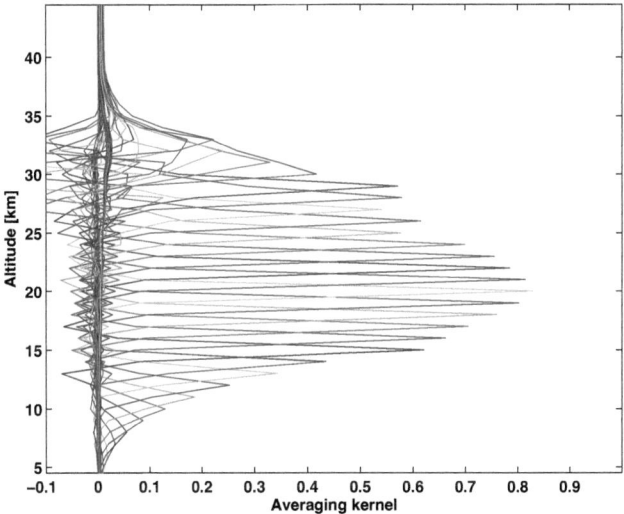

Figure 5.33: Averaging kernels of the second limb profile (indicated in Figure 5.32 by a black line) versus altitude.

116   CHAPTER 5. RESULTS AND DISCUSSION

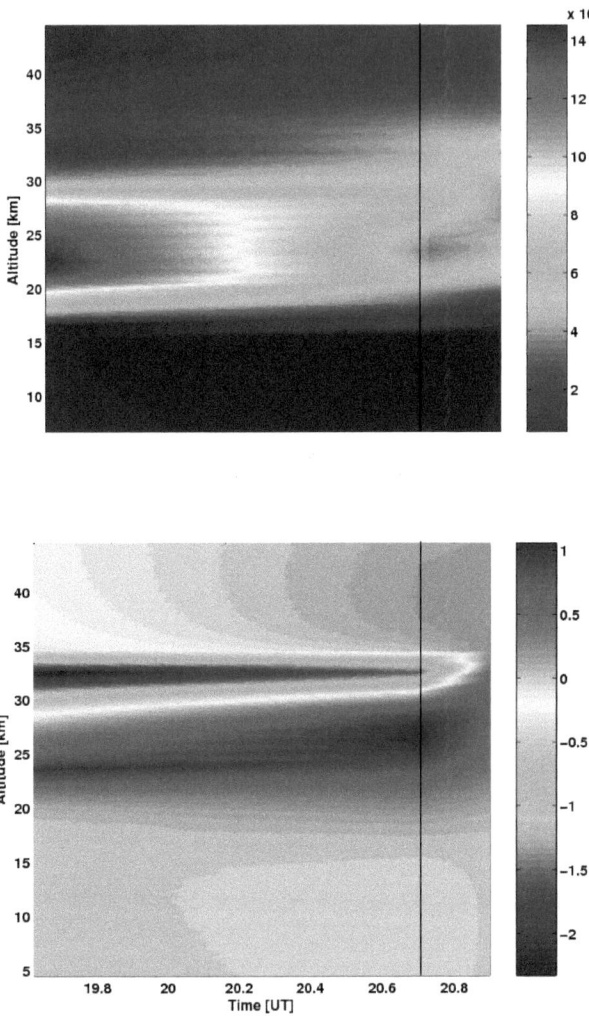

Figure 5.34: Retrieval of BrO from balloon-borne measurements on board LPMA/DOAS payload on June 17, 2005. Upper panel: BrO concentration versus altitude and time. Lower panel: Area of the averaging kernels versus altitude and time.

5.2. OBSERVATIONS FROM ABOARD LPMA/DOAS GONDOLA ON JUNE 17, 2005

Time series of BrO concentration profiles from LPMA/DOAS flight

As mentioned above, BrO measurements are only considered in the time frame from 19:36 UT to 20:54 UT and therefore only three profiles are retrieved.

The upper panel of Figure 5.34 shows the retrieved BrO concentration profiles versus time. A strong decrease is observed in the profiles from SZA = 75° to SZA = 90° (for SZA see Figure 5.22). While around 19:36 UT the concentration maximum exhibits around $1.4 \cdot 10^6$ molec/cm^3 the layer is hardly visible at around 20:46 UT and the concentration decreased to around $7 \cdot 10^6$ molec/cm^3.

The measurements yield the dominant contribution to the retrieved profiles between around 18 and 28 km altitude with increasing contribution towards lower SZA. Exemplarily, for the profile at 19:46 UT, which is designated in Figure 5.34 by a black line the averaging kernels are shown Figure 5.35. The indicated height resolution exhibits values of around 3 km between 18 and 28 km altitude. The retrieval of BrO from ascent and Limb scanning measurements from aboard LPMA/DOAS payload results in 10 degrees of freedom and a Shannon information content is 9.2 bits.

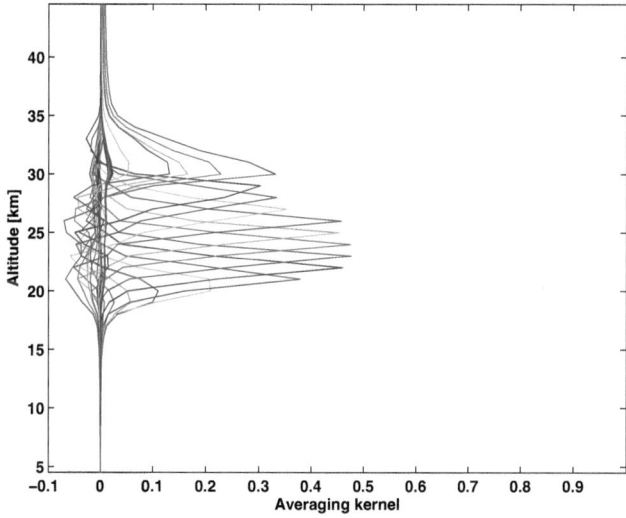

Figure 5.35: Averaging kernels of the second limb profile (indicated in Figure 5.34 by a black line) versus altitude.

5.2.5 Cross validation with direct sunlight DOAS measurements

In order to cross validate concentration profiles retrieved from scattered skylight measurements (mini-DOAS) and direct sunlight measurements (DOAS), their results from balloon ascent and sunset measurements of the same flight are compared with each other. In the case of O_3 in-situ sonde data is included.

Comparison with direct sun O_3

O_3 vertical profiles are compared in Figure 5.36. They agree within the error bars, only at 30 km the mini-DOAS result shows a peak in O_3, which is probably due to the above mentioned oscillation of the gondola at around 20:10 UT and propagates into the subsequent profile.

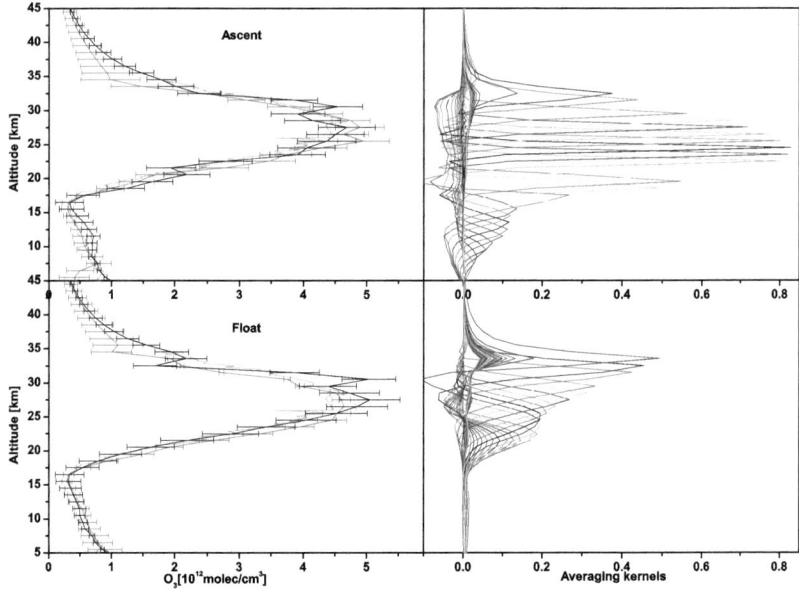

Figure 5.36: Left upper and lower panel: Concentration profile of O_3, from balloon-borne measurements of scattered light (black), direct sunlight (red) and from in-situ measurements by an electrochemical cell aboard the gondola (green). Right upper and lower panel: Averaging kernels of the scattered light retrieval.

5.2. OBSERVATIONS FROM ABOARD LPMA/DOAS GONDOLA ON JUNE 17, 2005

Comparison with direct sun NO_2

NO_2 vertical profiles are compared in Figure 5.37. For the ascent profile they agree in shape but the amount of NO_2 in the mini-DOAS profile is systematically lower compared to the direct sun profile. The profiles retrieved from measurements during balloon float (sunset) agree within the error bars, despite of the 25 to 28 km region, where the mini-DOAS results are larger than the direct sun results.

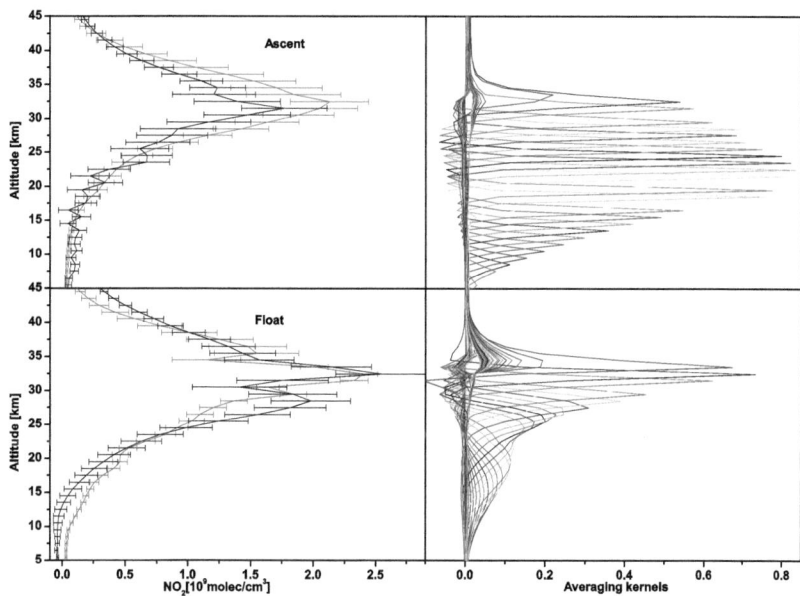

Figure 5.37: Left upper and lower panel: Concentration profile of NO_2 retrieved from balloon-borne measurements of scattered light (black) and direct sun (red). Right upper and lower panel: Averaging kernels of the scattered light retrieval.

Comparison with direct sun BrO

BrO vertical profiles are compared in Figure 5.38. From direct sun measuremts only an ascent profile is available. They agree within the error bars.

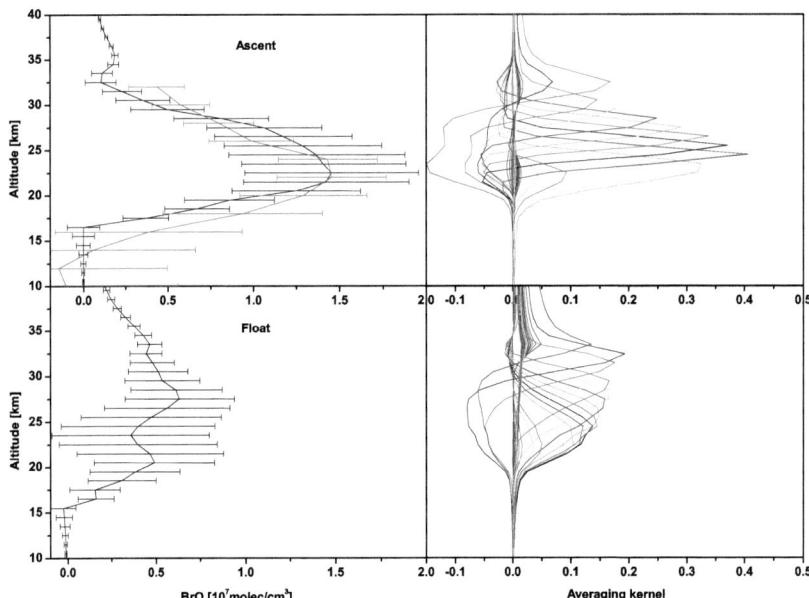

Figure 5.38: Left upper and lower panel: Concentration profile of BrO, retrieved from balloon-borne measurements of scattered light (black) and direct sun (red). Right upper and lower panel: Averaging kernels of the scattered light retrieval.

An overall good agreement is found between Limb scattered skylight and direct sunlight measurements, confirming the validity of the applied approach of even for strong concentration gradients in time.

5.3 Observations from aboard LPMA/IASI gondola on June 30, 2005.

On June 30 the mini-DOAS instrument was integrated on the LPMA-IASI (Infrared Atmospheric Sounding Interferometer) gondola, which is a modified version of the LPMA instrument used to observe the Earth's atmospheric emission at nadir, to determine the vertical profiles of various atmospheric gases such as H_2O, CO_2, CO, O_3, N_2O and CH_4 from high resolution atmospheric spectra.

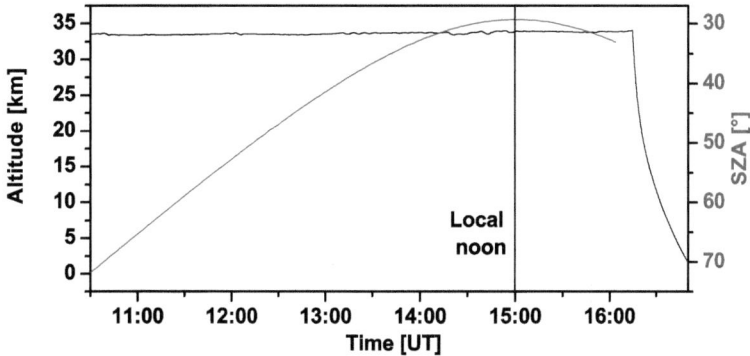

Figure 5.39: Flight profile of the IASI gondola on June 30, 2005. Altitude (black) and SZA (red) versus Universal Time.

5.3.1 Flight conditions

The payload was launched at around 6:30 UT, reaching float altitude of 33.4 km at around 9 UT. The gondola was cut at around 16:15 UT. The mini-DOAS instrument was measuring throughout that period. Due to problems in the RTM at high SZA data is shown here from 10:30 to 16 UT.

5.3.2 Measured ΔSCDs from LPMA/IASI flight

In a first step measured and forward modeled ΔSCDs are compared for the targeted gases (Figure 5.40–5.39). The wavelike structure in the ΔSCDs occurring for all gases is a consequence of the scanning telescope.

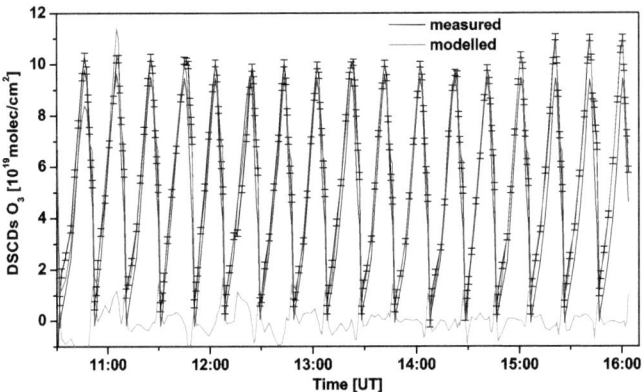

Figure 5.40: Measured (black) and forward modeled (red) O_3 ΔSCDs from Limb scanning measurements from aboard the LPMA/IASI gondola on June 30, 2005. The difference between forward modeled and measured ΔSCDs is shown in green.

ΔSCDs of O_3 from LPMA/IASI flight

Figure 5.40 shows measured and forward modeled O_3 ΔSCDs. The agreement is very good and the residual is mostly systematic. This can be due to a remaining pointing error, that seems to shift over the course of the measurements. The wavelike structure is due to the scanning telescope as described in detail for the flight on June 14.

5.3. OBSERVATIONS FROM ABOARD LPMA/IASI GONDOLA ON JUNE 30, 2005.

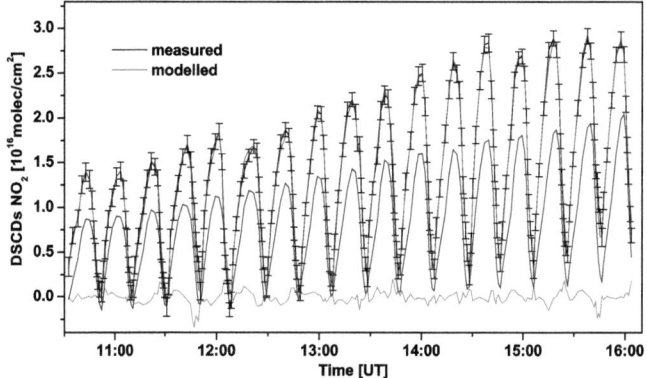

Figure 5.41: Measured (black) and forward modeled (red) NO_2 ΔSCDs from Limb scanning measurements from aboard the LPMA/IASI gondola on June 30, 2005. The difference between forward modeled and measured ΔSCDs is shown in green.

ΔSCDs of NO_2 from LPMA/IASI flight

Figure 5.41 shows measured and forward modeled NO_2 ΔSCDs. The agreement is very good and the residual is very low. Due to a constant balloon height, already the course of the ΔSCDs gives an indication of the increase of the NO_2 concentration during the day.

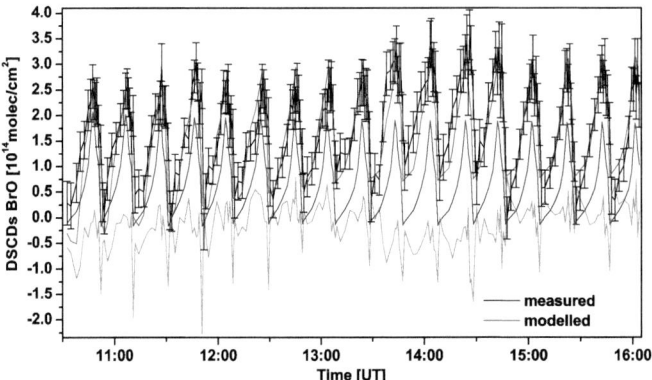

Figure 5.42: Measured (black) and forward modeled (red) BrO ΔSCDs from Limb scanning measurements from aboard the LPMA/IASI gondola on June 30, 2005. The difference between forward modeled and measured ΔSCDs is shown in green.

ΔSCDs of BrO from LPMA/IASI flight

Figure 5.42 shows measured and forward modeled BrO ΔSCDs. The residual contains random and systematic structures, but the comparison looks reasonable.

5.3. OBSERVATIONS FROM ABOARD LPMA/IASI GONDOLA ON JUNE 30, 2005.

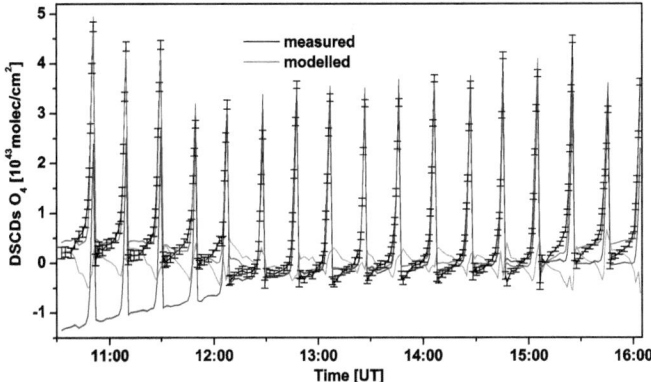

Figure 5.43: Measured (black) and forward modeled (red) O_4 ΔSCDs from Limb scanning measurements from aboard the LPMA/IASI gondola on June 30, 2005. The difference between forward modeled and measured ΔSCDs is shown in green.

ΔSCDs of O_4 from LPMA/IASI flight

Figure 5.43 shows measured and forward modeled O_4 ΔSCDs. There are remaining systematic structures in the residual and a change at around 12 UT and a smaller one at 15 UT can be observed.

5.3.3 Retrieved concentration profiles

In a second step the observations are inverted into trace gas concentration profiles, according to the method described in Section 4.4. The results of the inversion are shown in Figures 5.44 – 5.48. Each color plot is a composition of 12 height profiles. The upper panels on the left side show the retrieved trace gas concentration and the lower panels the area of the averaging kernels. The area of the averaging kernels gives an indication of the measurements contribution to the retrieved profiles (see section 4.4.5). The averaging kernels of the particular second limb profile (11:00 UT) are shown on the particular left side.

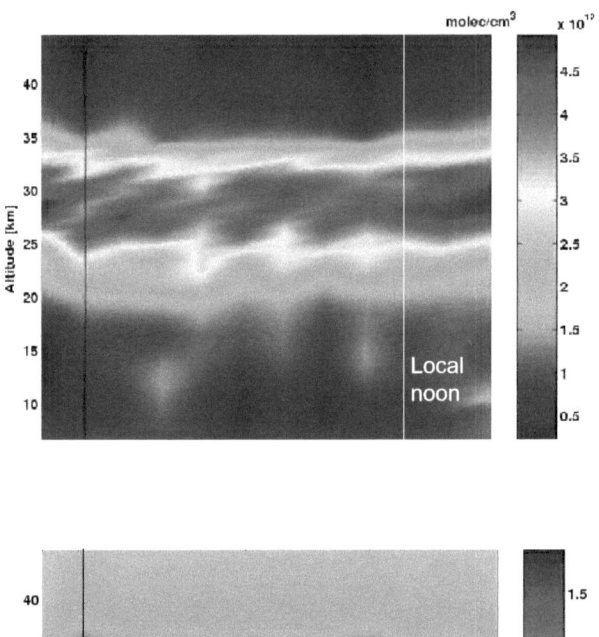

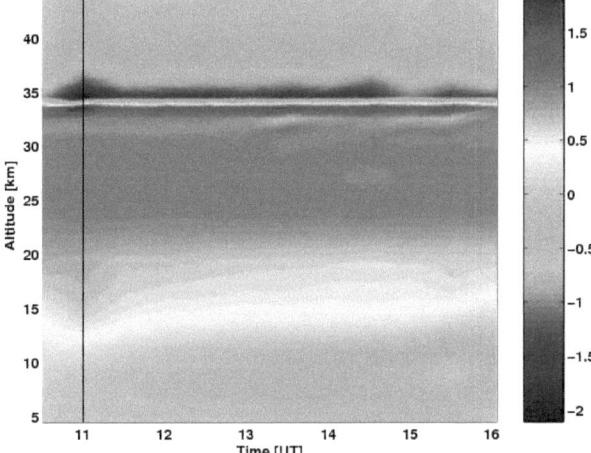

Figure 5.44: Retrieval of O_3 from balloon-borne measurements on board LPMA/IASI payload on June 14, 2005. Upper panel: O_3 concentration versus altitude and time. Lower panel: Area of the averaging kernels versus altitude and time.

5.3. OBSERVATIONS FROM ABOARD LPMA/IASI GONDOLA ON JUNE 30, 2005.

Time series of O_3 concentration profiles from LPMA/IASI flight

The upper panel of Figure 5.44 shows the retrieved O_3 concentration profiles versus time. The ozone concentration is highest in a layer between 24 and 33 km altitude, while a concentration maximum of around $5 \cdot 10^{12}$ molec/cm^3 is found at around 27 km altitude. The time variations are due to α oscillations of the gondola.

The area of the averaging kernels (lower panel) indicate a dominating contribution of the measurements to the results between 12 and 34 km. The upper limit of the main contribution by the measurements is given by the float height. The lower limit is due to the increasing importance of Rayleigh scattering in lower altitudes. Exemplarily, for the profile at 11:00 UT, which is designated in Figure 5.44 by a black line, the averaging kernels are shown Figure 5.45. They indicate a height resolution of around 1 km between 27 and 34 km altitude, decreasing slowly down to 7 km at 10 km altitude.

The retrieval of ozone from Limb scanning measurements from aboard LPMA/IASI payload results in 105 degrees of freedom and a Shannon information content is 198 bits.

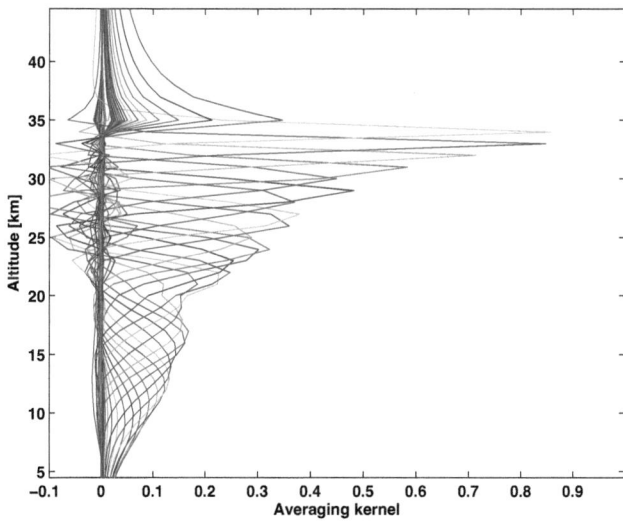

Figure 5.45: Averaging kernels of the second limb profile (indicated in Figure 5.44 by a black line) versus altitude.

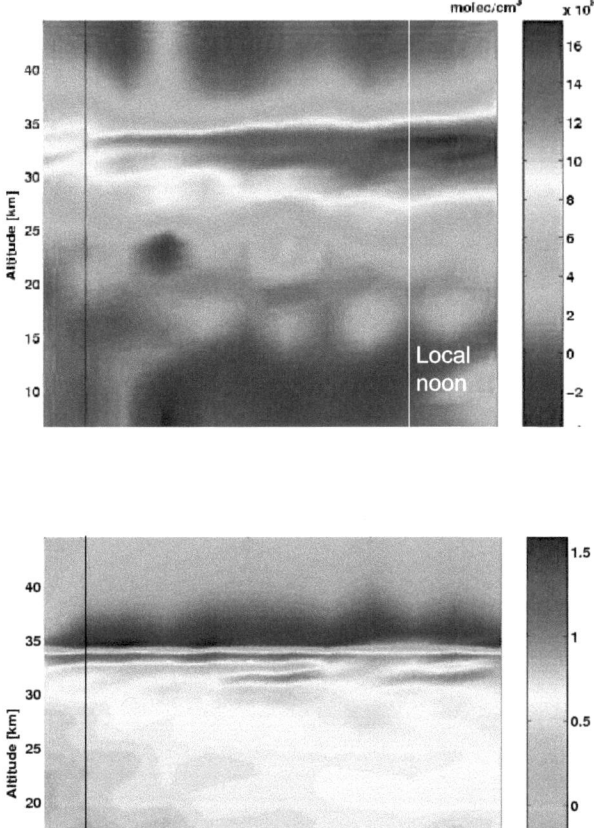

Figure 5.46: Retrieval of NO_2 from balloon-borne measurements on board LPMA/IASI payload on June 14, 2005. Upper panel: NO_2 concentration versus altitude and time. Lower panel: Area of the averaging kernels versus altitude and time.

5.3. OBSERVATIONS FROM ABOARD LPMA/IASI GONDOLA ON JUNE 30, 2005.

Time series of NO_2 concentration profiles from LPMA/IASI flight

The upper panel of Figure 5.46 shows the retrieved NO_2 concentration profiles versus time. The measurements of NO_2 show a very pronounced increase from around $9 \cdot 10^8$ molec/cm^3 at 10:30 UT to around $1.7 \cdot 10^9$ molec/cm^3 at 16 UT. The concentration of NO_2 is highest in a layer between 28 and 35 km altitude, while the concentration maximum is found at around 31 km altitude.

The measurements yield the dominant contribution to the retrieved profiles between around 15 and 34 km altitude. Exemplarily, for the profile at 11:00 UT, which is designated in Figure 5.46 by a black line, the averaging kernels are shown Figure 5.47. They indicate a height resolution of around 1 km between 27 and 34 km altitude, decreasing slowly down to 8 km at 12 km altitude. The retrieval of NO_2 from Limb scanning measurements from aboard LPMA/IASI payload results in 107 degrees of freedom and a Shannon information content is 199 bits.

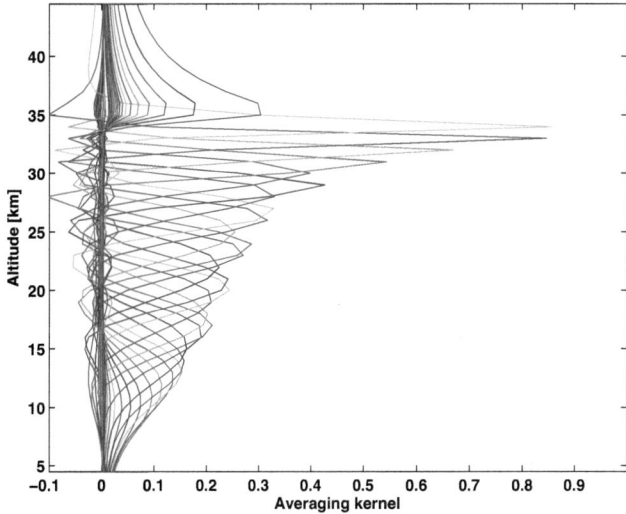

Figure 5.47: Averaging kernels of the second limb profile (indicated in Figure 5.46 by a black line) versus altitude.

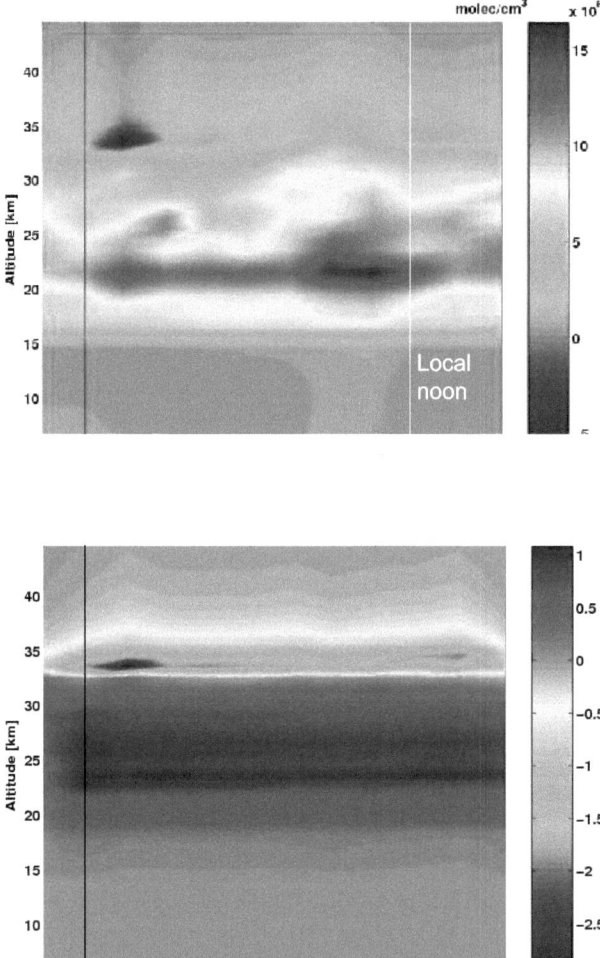

Figure 5.48: Retrieval of BrO from balloon-borne measurements on board LPMA/IASI payload on June 14, 2005. Upper panel: NO_2 concentration versus altitude and time. Lower panel: Area of the averaging kernels versus altitude and time.

5.3. OBSERVATIONS FROM ABOARD LPMA/IASI GONDOLA ON JUNE 30, 2005.

Time series of BrO concentration profiles from LPMA/IASI flight

The upper panel of Figure 5.48 shows the retrieved BrO concentration profiles versus time. The retrieved BrO concentration shows some variations but indicates a layer of highest values between 18 and 30 km altitude, while a concentration maximum of around $13 \cdot 10^6$ molec/cm^3 is found at around 23 km altitude. The measurements yield the dominant contribution to the retrieved profiles between around 20 and 33 km altitude. Exemplarily, for the profile at 11:00 UT, which is designated in Figure 5.48 by a black line, the averaging kernels are shown Figure 5.49. They indicate a height resolution of around 3 km between 23 and 34 km altitude. The retrieval of BrO from Limb scanning measurements from aboard LPMA/IASI payload results in 44 degrees of freedom and a Shannon information content is 41 bits.

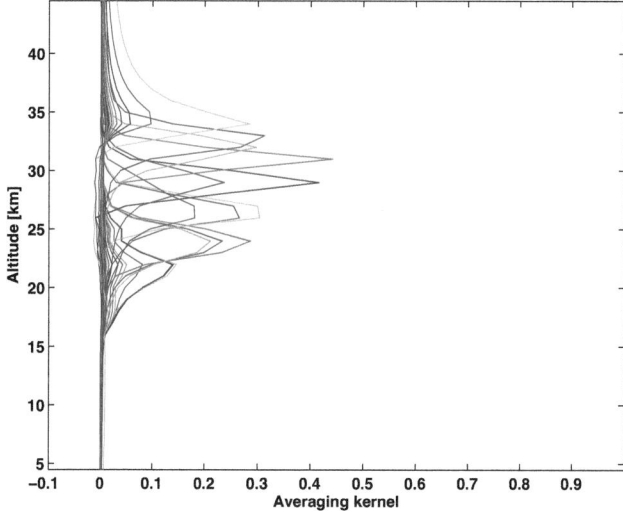

Figure 5.49: Averaging kernels of the second limb profile (indicated in Figure 5.48 by a black line) versus altitude.

5.3.4 Cross validation with satellite measurements

The main advantage of scattered skylight measurements compared to direct sun measurements is the possibility to derive time series of trace gas profiles during balloon float, independent of sunrise or sunset. Hence, we can compare profiles of radicals undergoing diurnal variation to satellite measurements without photochemical corrections, as is necessary for direct sun measurements (Butz, 2006; Dorf et al., 2006). The forward model is constructed in a way that the time grid of the retrieved profiles matches the time or actual SZA when the satellite instrument measured a profile at a location closest to the balloon borne measurements. This criterion is fulfilled for the overpass of SCIAMACHY/Envisat on June 30, 2005 at SZA = $34°$ for orbit 17427 with State ID 32 (Grunow, 2005). At 14 UT the same SZA was present at the position where the mini-DOAS instrument was measuring some 570 km away from the satellite pixel center.

The SCIAMACHY instrument is a UV/visible/near-IR spectrometer (220 nm - 2380 nm, FWHM: 0.2 nm - 1.5 nm), which was launched on sun-synchronous orbit on board the European Envisat satellite in March 2002. It is designed to measure in different viewing geometries, direct light from the sun or the moon in occultation mode, or sunlight scattered by the Earth's atmosphere in nadir or Limb direction (e.g.Burrows et al. (1995); Bovensmann (1999)). SCIAMACHY performs observations in Limb over an altitude range of 100 km, with a vertical step size of 3 km. Starting 3 km below the Earth horizon the atmosphere is scanned tangentially over a 1000 km wide swath. After each azimuth scan, the elevation is increased until the maximum altitude of 100 km is reached.

In the following profiles of NO_2, O_3 and BrO retrieved from SCIAMACHY measurements by the university of Bremen (Rozanov et al., 2005) are compared to our balloon-borne measurements. For NO_2 also a characterization of both retrievals is given. The comparison of the two profiles measured on different platforms is shown in Figure 5.50.

The quality of the retrievals is compared using the averaging kernel matrices shown in Figure 5.50, from which the area is derived as well as the degrees of freedom. In order to give a quantitative comparison of the height resolution of the profiles, the Backus-Gilbert spread (see section 4.4.5) is also calculated. For the satellite observations it indicates an altitude resolution of 2-4 km between 26 and 36 km decreasing to around 6 km below to 24 km and above to 38 km. For the balloon observations, the altitude resolution is 1 km between 32 and 36 km decreasing slowly down to 3 km at 14 km altitude. The area of the averaging kernels of the balloon borne retrieval is close to unity between 15 km and 35 km and for the satellite retrieval it is nearly 1 from 10 to 45 km altitude. The number of degrees of freedom for the satellite and balloon retrievals are 5.1 and 10.1, respectively.

Vertical stratospheric columns are derived from the retrieved profiles by summing up the concentrations from 18-40 km. According to the Bremen retrieval the vertical column is $1.51 \cdot 10^{15}$ molec/cm^2 and the balloon retrieval yields a vertical column of $1.36 \cdot 10^{15}$ molec/cm^2. The comparison of O_3 profiles is shown in Figure 5.51 (left panel). The SCIAMACHY profile is shifted in altitude compared to the balloon profile by around 2 km, which may be due to a pointing error in one of the retrievals. The BrO profile agree within their error bars, while the balloon profile indicates a slightly different shape.

An overall good agreement is found between balloon-borne and satellite measurements. The better height resolution below 25 km and the larger number of degrees of freedom qualifies the mini-DOAS measurements as appropriate tool for satellite validation/cross-comparison with the necessity of photochemical modeling.

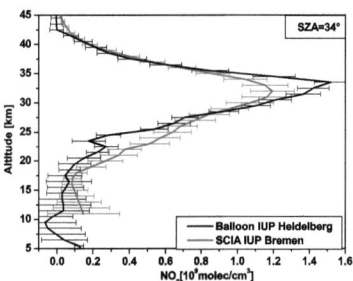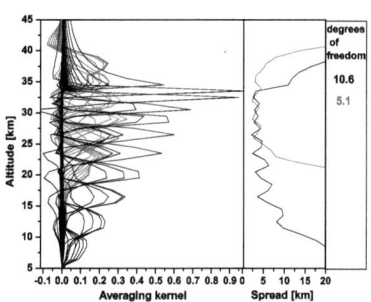

Figure 5.50: Left panel: Retrieved NO_2 concentration, from SCIAMACHY measurements in orbit 17427 by the IUP Bremen (red) and from balloon-borne measurements on board LPMA/IASI payload by the IUP Heidelberg (black), at SZA = 34° on June 30, 2005. Right panel: Left side: Averaging kernels of the retrieval of NO_2 from SCIAMACHY measurements by the IUP Bremen (red) and from balloon-borne measurements from the IUP Heidelberg (black). Middle: Backus-Gilbert spread of the averaging kernels. Right side: Degrees of freedom of the retrieval.

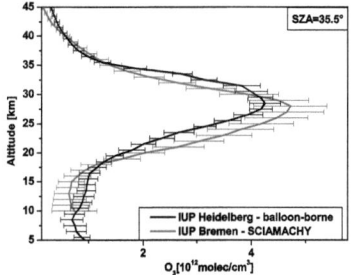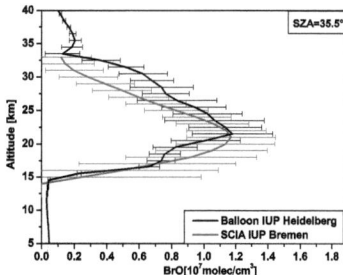

Figure 5.51: Retrieved concentration, from SCIAMACHY measurements in orbit 17427 by the IUP Bremen (red) and from balloon-borne measurements on board LPMA/IASI payload by the IUP Heidelberg (black), at SZA = 34° on June 30, 2005. Left panel: O_3. Right panel: BrO.

5.3.5 The diurnal variation of NO_2

The retrieved NO_2 concentration shows a nearly linear increase during the measurement period, which is explained by photochemical considerations in the following. While the overall reactive nitrogen content NO_y is invariant at short timescales, the partitioning changes rapidly around sunset and sunrise and slowly during day- and nighttime (Brasseur and Solomon, 2005). A reaction scheme including all reactions governing NO_y partitioning is displayed in Figure 5.52. The most important NO_y species in the

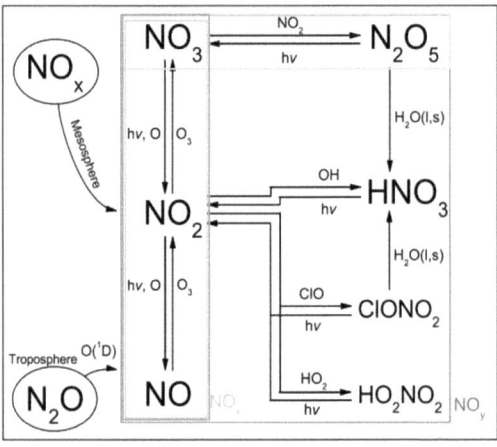

Figure 5.52: Reaction scheme of NO_y. The reactions in the orange boxes are considered in our simplified model.

tropical NO_x/NO_y layer (i.e. above 25 km) are N_2O_5 and HNO_3 at nighttime. This statement is well documented by the MIPAS-B measurements over Teresina, on June 14, 2005 (see Figure 5.54), which indicated at 30 km a concentration of about $1 \cdot 10^9$ molec/cm³ for HNO_3, $6 \cdot 10^8$ molec/cm³ for N_2O_5, $2.5 \cdot 10^8$ for $ClONO_2$ and $1.2 \cdot 10^9$ NO_2 at mid-night. Since the photolysis rate of HNO_3 is around one order of magnitude less than the photolysis rate of N_2O_5, the increase of NO_2 in the tropical daytime stratosphere is thus primarily due to the photolytic destruction of N_2O_5. The governing reactions are highlighted in Figure 5.52 by orange boxes. The monitoring of the diurnal increase of NO_2 (see chapter 1.8) may thus allow to infer the photolysis frequency of N_2O_5 from "in-situ" data.

5.3. OBSERVATIONS FROM ABOARD LPMA/IASI GONDOLA ON JUNE 30, 2005.

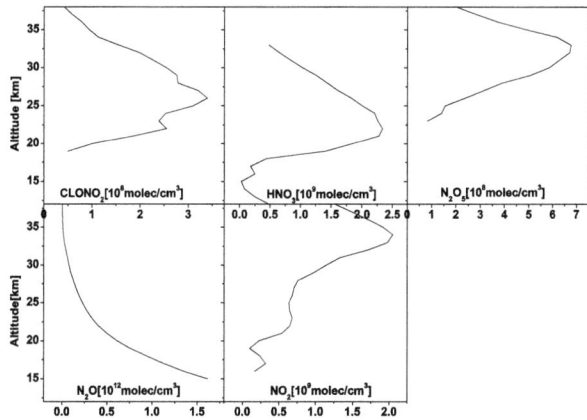

Figure 5.53: NO_x/NO_y species measured by MIPAS-B on June 14, 2005. Upper panels from left to right: $ClONO_2$, HNO_3 and N_2O_5. Lower panels: N_2O and NO_2.

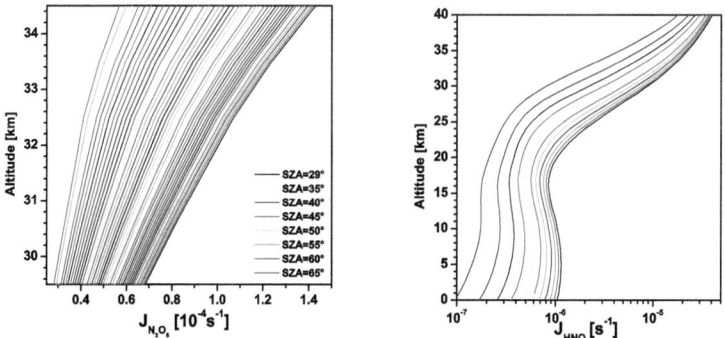

Figure 5.54: Photolysis rate of N_2O_5 ($J_{N_2O_5}$) for different SZAs (coloured lines), calculated by McArtim actinic fluxes using absorption cross sections from JPL 2006 (Sander et al., 2006).

136   CHAPTER 5. RESULTS AND DISCUSSION

The absorption cross sections of N_2O_5 were first measured by Jones and Wulf (1937). Since then, other measurements have been performed and significant temperature dependence at wavelength longer than about 280 nm were found (Brasseur and Solomon, 2005). In practice the governing equations are

$$2 \cdot \frac{d[N_2O_5]](SZA(t))}{dt} \approx \frac{d[NO_x]}{dt} \tag{5.4}$$

Using that $NO_3 \approx 0$ at daytime, due to rapid photolysis (life time around 10s (Sander et al., 2006)), NO_x in equation 5.4 can be inferred from the measured NO_2 and the established correlation with O_3 (also measured by the mini-DOAS instrument) and the rate constant k from reaction R1.35 (taken from JPL2006 (Sander et al., 2006)).

$$[NO] = \frac{J_{NO_2} \cdot [NO_2]}{k \cdot [O_3]}. \tag{5.5}$$

or

$$\frac{d[NO_x(t)]}{dt} = \frac{d([NO]+[NO_2])}{dt} \approx (1 + \frac{J_{NO_2}}{k \cdot [O_3]}) \cdot \frac{[NO2(t)]}{dt}). \tag{5.6}$$

Here J_{NO_2} is assumed constant over the considered SZA range from 30 to 70°, due to the cross section's maximum around 400 nm, where the actinic flux's change is in the order of 5%. Inserting equation 5.6 into 5.4 leads to

$$2 \cdot \frac{d[N_2O_5]}{dt} \approx \frac{d[NO_2]}{dt} \cdot (1 + \frac{J_{NO_2}}{k \cdot [O_3]}) \tag{5.7}$$

As discussed above, the decrease of N_2O_5 is primarily due to the photolysis rate $J_{N_2O_5}$,

$$\frac{d[N_2O_5]}{dt} \approx -J_{N_2O_5}(t) \cdot [N_2O_5](t) \tag{5.8}$$

in the our comparison study, equation 5.8 is integrated in steps of 1° SZA.

$$[N_2O_5](t) = [N_2O_5](t_0) \cdot \exp(-J_{N_2O_5}(t) \cdot t) \tag{5.9}$$

Inserting equation 5.9 into equation 5.8 and the resulting equation into equation 5.7 leads to

$$\frac{d[NO_2]}{dt} = \frac{2 \cdot J_{N_2O_5}(t) \cdot [N_2O_5](t_0) \cdot \exp(-J_{N_2O_5}(t) \cdot t)}{(1 + \frac{J_{NO_2}}{k \cdot [O_3]})} \tag{5.10}$$

Finally the integration of equation 5.10 from t_0 to t leads to

$$[NO_2](t) = [NO_2](t_0) + \frac{2 \cdot J_{N_2O_5}(t)[N_2O_5](t_0) \cdot \exp(-J_{N_2O_5}(t) \cdot t) \cdot t}{(1 + \frac{J_{NO_2}}{k \cdot [O_3]})} \tag{5.11}$$

which gives us the model function for the increase of NO_2 during the measurement period.
The photolysis frequency $J_{N_2O_5}$ is obtained using the following data:
The absorption cross-section and quantum yield for N_2O_5 are available from laboratory studies for a temperature range of 233 to 295 K (Sander et al., 2006). For the present study the temperature dependent parameterization of the absorption cross-section is interpolated to the required temperature range. At 23 km T was 210 K and therefore below the limit of the given range (Figure 5.2). N_2O_5 absorption cross sections show no strong temperature dependence at wavelength smaller than 250 nm. Temperature dependence becomes significant at wavelengths above 280 nm, where cross sections decrease with decreasing temperature, and the effect increases with increasing wavelength. Combined uncertainty for

5.3. OBSERVATIONS FROM ABOARD LPMA/IASI GONDOLA ON JUNE 30, 2005.

cross section and quantum yield are a factor of 2, referring to the total dissociation rate. This factor gives the uncertainty range for the modeled increase of NO_2.
The actinic fluxes are modeled using the radiative transfer model McArtim for the relevant SZA range in steps of 1° and wavelength range of 200 to 410 nm. The resulting photolysis rates (for the calculation see 2.1) are shown in Figure 5.54. Since $J_{N_2O_5}$ changes with SZA and SZA changes with time, the calculation of $[NO_2](t)$ (5.11) is performed stepwise (in steps of 1° ΔSZA), while $J_{N_2O_5}$ is assumed to be constant during each step. $[N_2O_5](t_0)$ is taken from a Labmos model run with an initialization value of

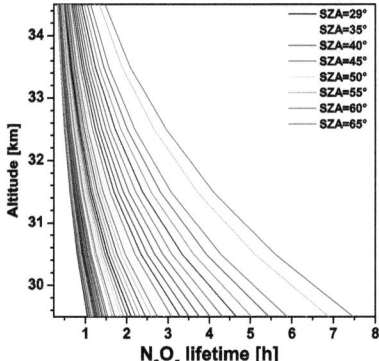

Figure 5.55: Inverse lifetime of N_2O_5 due to photolysis for an SZA range from 72 to 29°, using the molecular and kinetic data from JPL 2006 (Sander et al., 2006).

N_2O_5 measured by MIPAS-B over Teresina on June 14, 2005 [personal communication with G. Wetzel, IMK Karlsruhe]. The resulting time series of NO_2 concentration is displayed in Figure 5.56 for different altitudes. The measured increase of NO_2 is slightly larger than the modeled increase, but agrees within the given error bars.

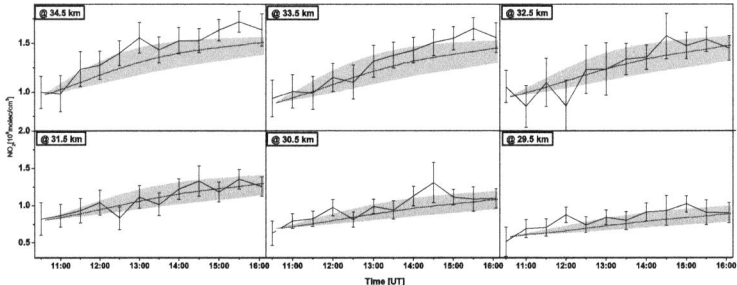

Figure 5.56: Time series of measured (black) and modeled (gray) NO_2 concentration for different altitudes. The uncertainty range of the modeled data is due to the combined uncertainty for cross section and quantum yield.

Conclusion

Summary

The present study addresses the retrieval and interpretation of time dependent trace gas profiles from balloon-borne limb scattered skylight measurements. The measurements were performed during three deployments of the mini-DOAS instrument on different balloon payloads at a tropical station in northern Brazil.

By applying the DOAS technique, the recorded spectra were analyzed for O_3, NO_2, BrO, O_4 and HONO signatures. For the interpretation of the retrieved trace gas slant column densities the radiative transfer was modeled using the RTM McArtim, which provided the weighting functions in altitude space and limb radiances of the different viewing geometries as an output.

In order to improve the determination of the viewing geometry, the elevation of the telescope (α) was retrieved by the comparison of measured and modeled relative radiances at different wavelengths.

Detailed sensitivity studies were performed in order to estimate the error produced by pendulum oscillations of the gondola. These studies showed, that the oscillations of the gondola are potentially large error sources in the retrieved profiles. This effect increases with the spatial distance between the observer and the object, i.e. for the presented balloon floats at around 34 km altitude the resulting error caused by oscillations is highest for BrO and O_3 and lower for NO_2. The error arising from these oscillations can not be generalized, here 10%, 7.5% and 5% are considered for BrO, O_3 and NO_2, respectively.

A new retrieval algorithm was implemented in order to retrieve time series of upper tropospheric and lower and middle stratospheric trace gas profiles. Since the weighting function formalism considered the relative character of the measurements, the balloon borne mini-DOAS measurements are in that respect self calibrated, i.e. no information on overhead absorbers is necessary. The algorithm took into account the temporal distance between the measurement and the state and is, therefore, well suited for the retrieval of the diurnal variation of UV/vis absorbing radicals. Since the forward model requires no chemical modeling as an input, the retrieval method provides a tool for testing photochemical parameters. Different inversion techniques with respect to additional constraints were tested. Since the inversion of limb scattered measurements to trace gas profiles is formally ill-posed, the retrieval had to be constrained. Here the optimal estimation method was chosen, allowing for a detailed characterization of the retrieved profiles.

CONCLUSION

The height resolution of the mini-DOAS measurements ranges from 1 km between 32 and 36 km altitude, decreasing slowly down to 3 km at 14 km altitude. The number of independent parameters in a single retrieved profile depend on the applied time spacing and achieve e.g. 10.1 in the case of NO_2. The retrieval on an half hour time grid, from 10 to 16 UT, results in 101 degrees of freedom.

Several comparison studies were performed, where trace gas profiles retrieved from the mini-DOAS measurements were compared to in-situ ozone sonde data, satellite observations from Envisat/SCIAMACHY and direct sun measurements. Different approaches were chosen for the comparison of different measurement techniques performed on different platforms. For the comparison with in-situ ozone data, the high resolution profile from sonding was degraded into the altitude resolution of our measurements. Both SCIAMACHY and direct sun trace gas profiles (O_3, NO_2, BrO) were compared to our measurements retaining their inherent altitude resolution, which was well documented in the comparison of the averaging kernels. These comparison studies demonstrated the strength and validity of our approach which renders meteorological and photochemical corrections of measured radical concentrations due to temporal mismatches of corresponding observations unnecessary.

A tropospheric aerosol extinction profile could be retrieved from the comparison of measured and modeled O_4 slant column densities and agrees well with the results of retrieving the aerosol extinction profile from measured relative radiances.

The possibility of the observation of airmasses at changing illumination extends the possibilities of satellite, aircraft, in-situ and direct sun measurements and supports further applications. Two studies on photochemistry were performed within this thesis. One is the attempt to explain the observation of HONO by contemporary lightning and established photochemistry. The other study aims at drawing a conclusion on the photolysis rate of N_2O_5 (in-situ).

The explanation of the detection of HONO in the upper troposphere nearby thunderstorms, on the background of NO_x produced by lightning, requires an unknown OH source or additional reaction forming HONO.

For the interpretation of the diurnal variation of NO_2, a simplified model of the time dependent variation of NO_2 as a result of N_2O_5 photolysis is implemented. The model includes NO_y species measured by the MIPAS-B instrument and O_3 from mini-DOAS measurements. The N_2O_5 photolysis rates are derived from RTM actinic fluxes and absorption cross sections from JPL-2006. The comparison of modeled and measured NO_2 increase indicates a slightly larger N_2O_5 photolysis rate than the one in JPL-2006, but agrees within the given error bars of a factor of 2.

Outlook

The presented retrieval algorithm of retrieving time dependent trace gas profiles provides a tool for the investigation of a variety of questions regarding photochemistry.
Monitoring the diurnal variation of stratospheric radicals offers field tests concerning crucial photo-

CONCLUSION

chemical parameters important for stratospheric ozone, such as the photolysis frequencies of $BrONO_2$ or the efficiency of the ClO/BrO ozone loss cycle by simultaneous observations of the diurnal variation of NO_2, BrO and OClO. However, some technical advancements in the instrumental set up as well as in the retrieval could largely improve the quality of the results.

A coupled model of radiative transfer and photochemistry would provide weighting functions for photochemical parameters, which then could be directly retrieved from a set of slant column densities.

In order to avoid complications due to the uncertainties in the elevation of the telescope, the application of additional sensors providing the detector's elevation would increase the accuracy of the retrieved vertical profiles. In an autonomous approach, the mini-DOAS instrument would hold its own attitude control system, but depending on the balloon payload the mini-DOAS instrument is housing, the payloads attitude control system could be used and data transfered to the mini-DOAS computer and integrated in the measurement routine.

When measuring also in limb geometry, e.g. the MIPAS-B instrument, which is equipped with a precise pointing system, the viewing geometries could be directly connected to each other.

Instruments, which are similar to the mini-DOAS, are recently deployed for aircraft applications and will be in future also with the extension on the infrared spectral range. Even when employing different viewing geometries and spectral retrieval techniques, the presented algorithm for retrieving time dependent trace gas profile can be adapted to a variety of different applications.

Appendix

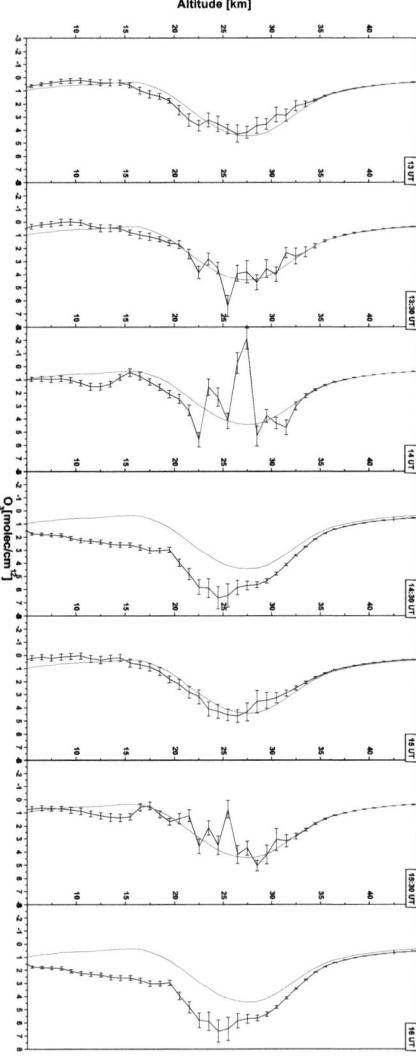

Figure 8.1: Subsequent concentration profiles of O_3, retrieved from mini-DOAS measurements on the Mipas-B payload on June 14, 2005

APPENDIX

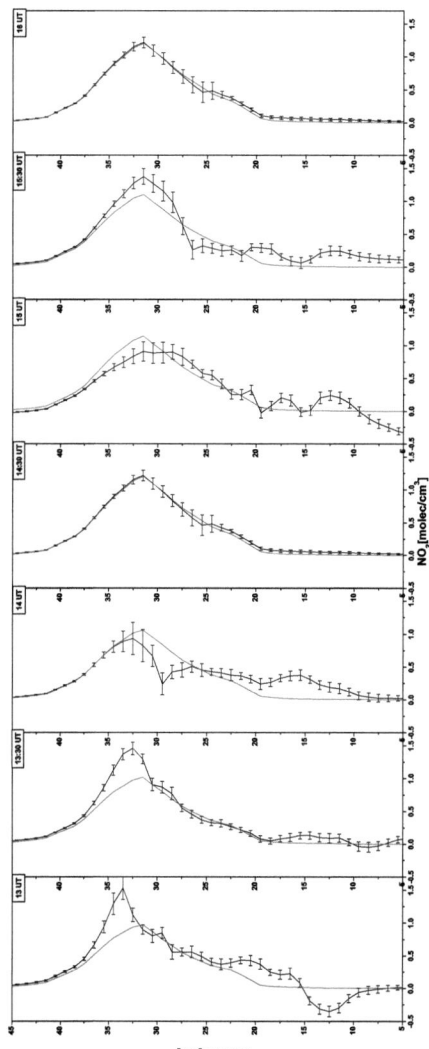

Figure 8.2: Subsequent concentration profiles of NO_2, retrieved from mini-DOAS measurements on the Mipas-B payload on June 14, 2005

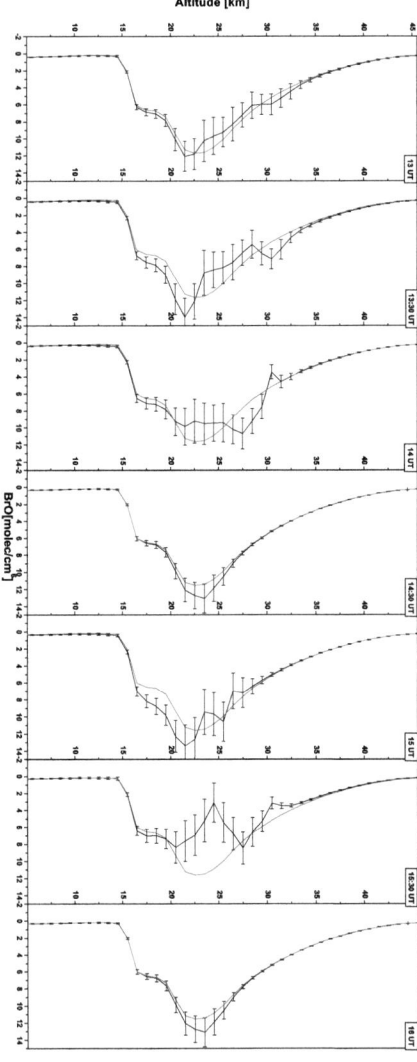

Figure 8.3: Subsequent concentration profiles of BrO, retrieved from mini-DOAS measurements on the Mipas-B payload on June 14, 2005

APPENDIX

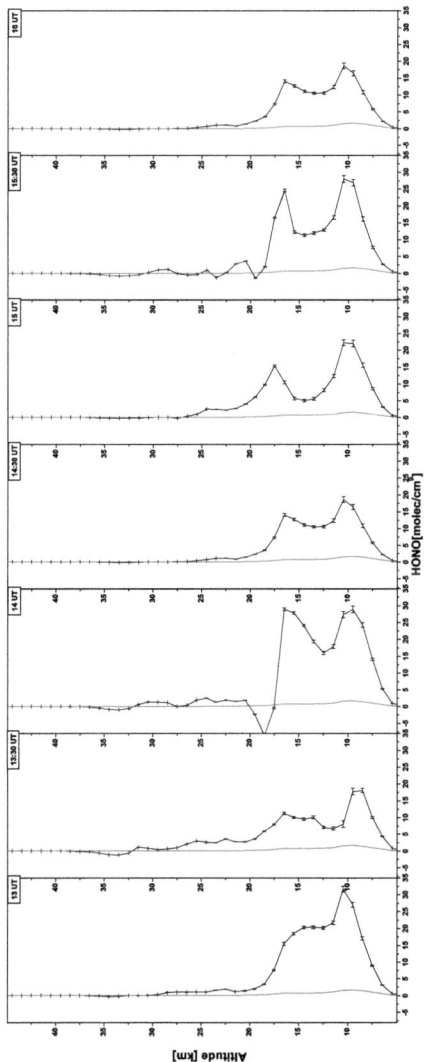

Figure 8.4: Subsequent concentration profiles of HONO, retrieved from mini-DOAS measurements on the Mipas-B payload on June 14, 2005

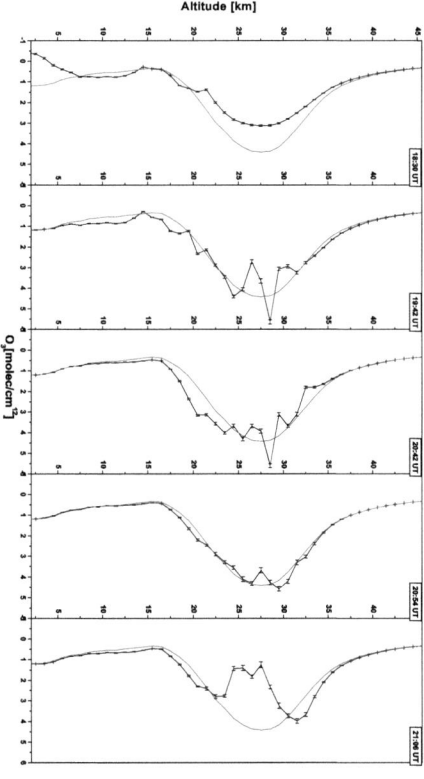

Figure 8.5: Subsequent concentration profiles of O_3, retrieved from mini-DOAS measurements on the LPMA/DOAS payload on June 17, 2005

APPENDIX

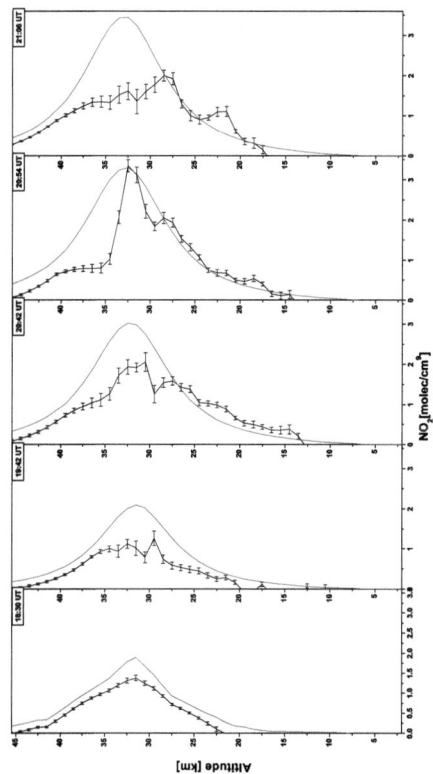

Figure 8.6: Subsequent concentration profiles of NO_2, retrieved from mini-DOAS measurements on the LPMA/DOAS payload on June 17, 2005

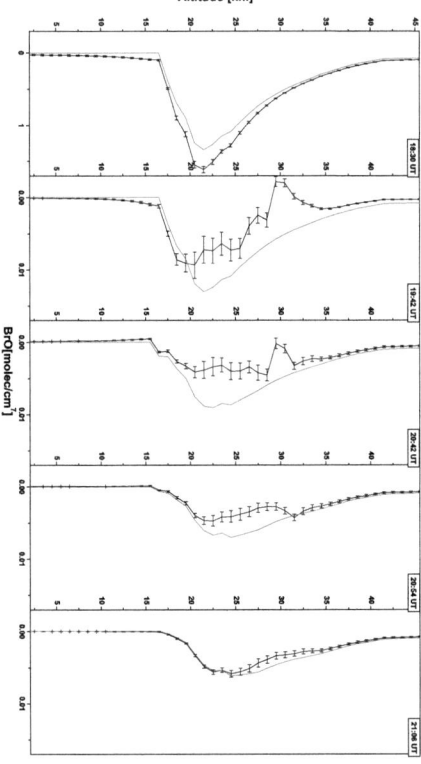

Figure 8.7: Subsequent concentration profiles of BrO, retrieved from mini-DOAS measurements on the LPMA/DOAS payload on June 17, 2005

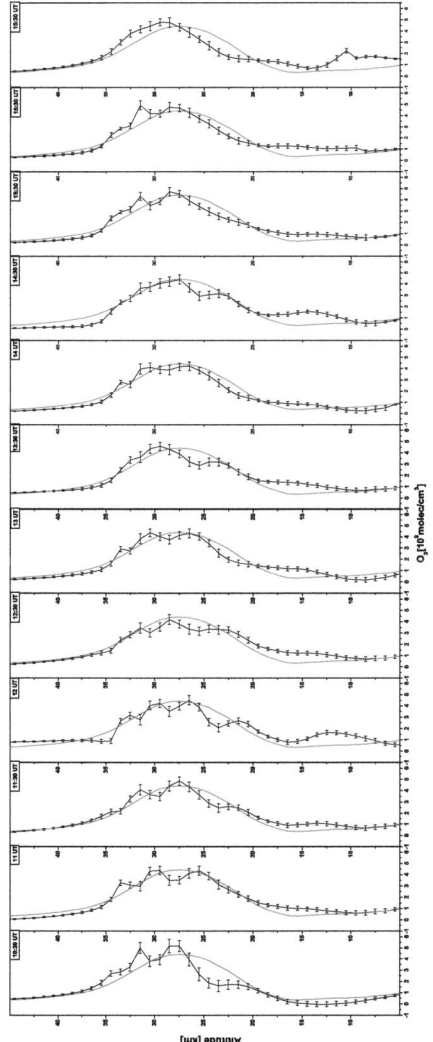

Figure 8.8: Concentration profile of O_3, retrieved from balloon-borne measurements on board LPMA/IASI payload on June 14, 2005 (black) and a *priori* profile (red).

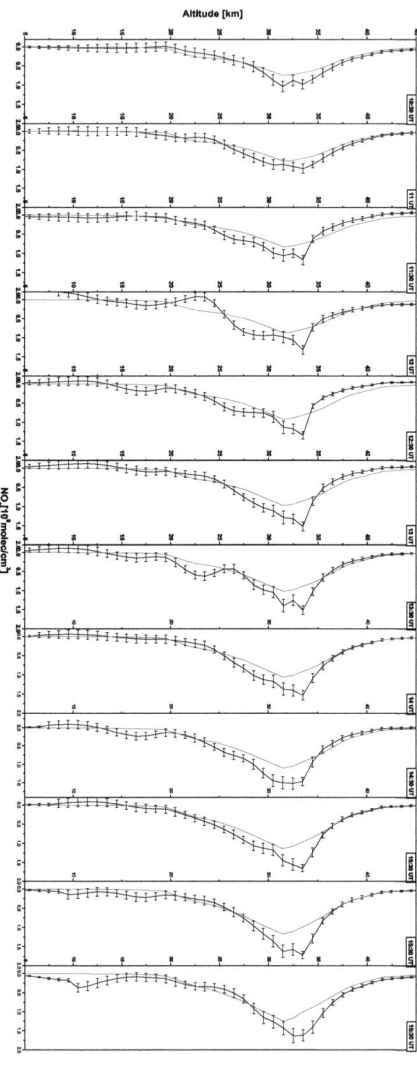

Figure 8.9: Concentration profile of NO_2, retrieved from balloon-borne measurements on board LPMA/IASI payload on June 14, 2005 (black) and *a priori* profile (red).

APPENDIX

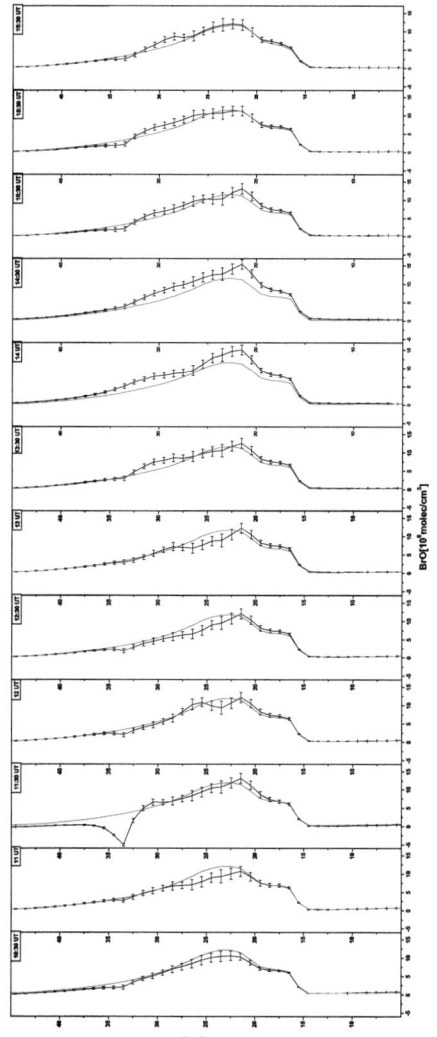

Figure 8.10: Concentration profile of BrO, retrieved from balloon-borne measurements on board LPMA/IASI payload on June 14, 2005 (black) and *a priori* profile (red).

Bibliography

Alicke, B., Geyer, A., Hofzumahaus, A., Holland, F., Konrad, S., H.-W. Paetz, J. S., Stutz, J., Volz-Thomas, A., and Platt, U.: OH formation by HONO photolysis during the BERLIOZ experiment, J. Geophys. Res., 108, 8247, 2003.

Alliwell, S. R., Van Roozendael, M., Johnston, P. V., Richter, A., Roozendael, M. V., Wagner, T., W.Arlander, D., Burrows, J. P., Fish, D. J., Jones, R. L., Tørnkvist, K. K., Lambert, J. C., Pfeilsticker, K., and Pundt, I.: Analysis for BrO in zenith-sky spectra - an intercomparison exercise for analysis improvement, J. Geophys. Res., 2002.

Backus, G.E.and Gilbert, J.: Uniqueness in the inversion of inaccurate gross earth data, 1970.

Bates, D. R. and Nicolet, M.: Atmospheric hydrogen, Publ. Astron. Soc. Pac., 62, 106–110, 1950.

Bertram, T. H., Perring, A. E., Wooldridge, P. J., Crounse, J. D., Kwan, A. J., Wennberg, P. O., Scheuer, E., Dibb, J., Avery, M., Sachse, G., Vay, S. A., Crawford, J. H., McNaughton, C. S., Clarke, A., Pickering, K. E., Fuelberg, H., Huey, G., Blake, D. R., Singh, H. B., Hall, S. R., Shetter, R. E., Fried, A., Heikes, B. G., and Cohen, R. C.: Direct Measurements of the Convective Recycling of the upper Troposphere, Science, 315, 10.1029/2002JD002299, 2007.

Birk, M., Mair, U., Krocka, M., Wagner, G., Yagoubov, P., Hoogeveen, R., Graauw, T. d., Stadt, H. v. d., Selig, A., Habers, H.-W., Richter, H., Semenov, A., Koshelets, V., Shitov, S., Ellison, B., Matheson, D., Alderman, B., Harman, M., Kerridge, B., Siddans, R., Reburn, J., Duric, A., Murk, A., Magun, A., Kämpfer, N., and Murtagh, D.: TELIS development of a new balloon borne THz/submmW heterodyne limb sounder, Final Programme, pp. 352 – 352, COSPAR, 35th COSPAR Scientific Assembly, 35th COSPAR Scientific Assembly, Paris, France, abstract on accompanying CD.

Bovensmann, H.: SCIAMACHY: Mission Objectives and Measurement Modes, J. Atmos. Sci., 56, 127–150, 1999.

Bovensmann, H., Burrows, J., Buchwitz, M., Frerick, J., Noel, S., Rozanov, V., Chance, K., and Goede, A.: SCIAMACHY: Mission Objectives and Measurement Modes, J. Atmos. Sci., 56, 127–150, 1999.

Brasseur, G. and Solomon, S.: Aeronomy of the middle atmosphere, D. Reidel Publ., Dordrecht, Boston, Lancaster, Tokyo, 1986.

Brasseur, G. and Solomon, S.: Aeronomy of the middle atmosphere, Springer, P.OBox 17, 3300 AADordrecht, The Netherlands, 2005.

Burrows, J., Hölzle, E., Goede, A., Visser, H., and Fricke, W.: SCIAMACHY - Scanning Imaging Absorption Spectrometer for Atmospheric Chartography, Acta Astronautica, 35, 445, 1995.

Bussemer, M.: Der Ring-Effekt: Ursachen und Einfluß auf die spektroskopische Messung stratosphärischer Spurenstoffe, Diploma thesis, Institut für Umweltphysik, Universität Heidelberg, 1993.

Butz, A.: Case Studies of Stratospheric Nitrogen, Chlorine and Iodine Photochemistry Based on Balloon Borne UV/visible and IR Absorption Spectroscopy, Phd thesis, Heidelberg University, Germany, 2006.

Camy-Peyret, C., Jeseck, P., Hawat, T., Durry, G., Payan, S., Berube, G., Rochette, L., and Huguenin, D.: The LPMA Balloon-Borne FTIR Spectrometer: Remote Sensing of Atmospheric Constituents, Proceedings of the 12th ESA Symposium on Rocket and Balloon Programmes and Related Research, Lillehammer, Norway, pp. 323–328, 1995.

Chapman, S.: On ozone and atomic oxygen in the upper atmosphere, Philos. Mag., 10, 369–383, 1930.

Connor, B. J., Siskind, D. E., Tsou, J. J.and Parrish, A., and Remsberg, E. E.: Ground-based microwave observations of ozone in the upper stratosphere and mesosphere, J. Geophys. Res., 99 (D8), 16 757–16 770, 1994.

Cox, R.: Photolysis of gaseous nitrous acid, Journal of Photochemistry, 3, 175 188, 1974.

Crutzen, P. J.: The influence of nitrogen oxide on the atmospheric ozone content, Q.J.R. Meteorol. Soc., 96, 320–327, 1970.

Demtröder, W.: Experimentalphysik 3 - Atome, Moleküle und Festkörper, Springer, Berlin, 2. edn., 2000.

Deutschmann, T.: Atmospheric Radiative Transfer Modelling with Monte Carlo Methods, Diploma thesis, Institut für Umweltphysik, Universität Heidelberg, 2008.

Dix, B., Brenninkmeijer, C. A. M., Frieß, U., Wagner, T., and Platt, U.: Airborne multi-axis DOAS measurements of atmospheric trace gases on CARIBIC long-distance flights, Atmospheric Measurement Techniques Discussions, 2, 265–301, http://www.atmos-meas-tech-discuss.net/2/265/2009/, 2009.

Dorf, M.: Investigation of Inorganic Stratospheric Bromine using Balloon-Borne DOAS Measurements and Model Simulations, Ph.D. thesis, Institut fuer Umweltphysik, University of Heidelberg, 2005.

Dorf, M., Bösch, H., Butz, A., Camy-Peyret, C., P., C. M., Engel, A., Goutail, F., Grunow, K., Hendrick, F., Hrechanyy, S., Naujokat, B., Pommereau, J.-P., Van Roozendael, M., Sioris, C., Stroh, F., Weidner, F., and Pfeilsticker, K.: Balloon-borne stratospheric BrO measurements: comparison with Envisat/SCIAMACHY BrO limb profiles, Atmos. Chem. Phys., 6, 2483–2501, 2006.

Farman, J., Gardiner, B., and Shanklin, J.: Large losses of total ozone in Antarctica reveal seasonal ClO_x/NO_x interaction, Nature, 315, 207–210, 1985.

Fayt, C. and van Roozendael, M.: WinDOAS 2.1. Software User Manual, technical report, see http://www.oma.be/BIRA-IASB/Molecules/BrO/WinDOAS-SUM-210b.pdf, 2001.

Ferlemann, F.: Ballongestützte Messung stratosphärischer Spurengase mit differentieller optischer Absorptionsspektroskopie, Ph.D. thesis, Institut für Umweltphysik, University of Heidelberg, 1998.

Ferlemann, F., Camy-Peyret, C., Fitzenberger, R., Harder, H., Hawat, T., Osterkamp, H., Schneider, M., Perner, D., Platt, U., Vradelis, P., and Pfeilsticker, K.: Stratospheric BrO profiles measured at different latitudes and seasons: Instrument description, spectral analysis and profile retrieval, Geophys. Res. Lett., 25, 3847–3850, 1998.

Ferlemann, F., Bauer, N., Harder, H., Osterkamp, H., Perner, D., Platt, U., Schneider, M., Vradelis, P., and Pfeilsticker, K.: A new DOAS-instrument for stratospheric balloon-borne trace gas studies, Appl. Opt., 39, 2377–2386, 2000.

Fleischmann, O., Hartmann, M., Orphal, J., and Burrows, J.: UV absorption cross sections of BrO for stratospheric temperatures (203-293K) recorded by a Time-resolved Rapid Scan FTS method, for details see: http://www.iup.physik.uni-bremen.de/gruppen/molspec/bro2_page.html, 2000.

Friess, U., Monks, P. S., Remedios, J. J., Rozanov, A., Sinreich, R., Wagner, T., , and Platt, U.: MAX-DOAS O4 measurements: A new technique to derive information on atmospheric aerosols: 2. Modeling studies, J. Geophys. Res., 2006.

Fueglistaler, S., Dessler, A. E., Dunkerton, T. J., Folkins, I., Fu, Q., and Mote, P. W.: Tropical tropopause layer, REVIEWS OF GEOPHYSICS, 47, 2009.

Goody, R. M. and Yung, Y. L.: Atmospheric Radiation: Theoretical Basis, Oxford University Press, 1989.

Grainger, J. and Ring, J.: Anomalous Fraunhofer line profiles, Nature, 193, 762, 1962.

Grunow, K.: Matching aircraft and balloon-borne measurements with Envisat observations: Comparison of two trajectory calculation tools, Atmospheric Chemistry and Physics, 2005.

Harder, H., Camy-Peyret, C., Ferlemann, F., Fitzenberger, R., Hawat, T., Osterkamp, H., Schneider, M., Perner, D., Platt, U., Vradelis, P., and Pfeilsticker, K.: Stratospheric BrO profiles measured at different latitudes and seasons: Atmospheric observations, Geophys. Res. Lett., 25, 3843–3846, 1998.

Harder, J., Brault, J., Johnston, P., and Mount, G.: Temperature dependent NO_2 Cross Section at high Spectral Resolution, J. Geophys. Res., 102, 3861–3879, 1997.

Hendrick, F., Barret, B., Roozendael, V., M., Bösch, H., Butz, A., De Maziere, M., Goutail, F., Hermans, C., Lambert, J.-C., Pfeilsticker, K., and Pommereau, J.-P.: Retrieval of nitrogen dioxide stratospheric profiles from ground-based zenith-sky UV-visible observations: validation of the technique through correlative comparisons, Atmos. Chem. Phys., 4, 2091–2106, 2004a.

Hendrick, F., Barret, B., Van Roozendael, M., Bösch, H., Butz, A., De Maziere, M., Goutail, F., Hermans, C., Lambert, J.-C., Pfeilsticker, K., and Pommereau, J.-P.: Retrieval of nitrogen dioxide stratospheric profiles from ground-based zenith-sky UV-visible observations: validation of the technique through correlative comparisons, Atmos. Chem. Phys., 4, SRef-ID: 1680-7324/acp/2004-4-2091, 2091–2106, 2004b.

Hermans: Private communication, for details see: http://www.oma.be/BIRA-IASB/Scientific/Topics/lower/LaboBase/acvd/O4Info.html, 2002.

Huntrieser, H., Schumann, U., Schlager, H., Höller, H., Giez, A., Betz, H.-D., Brunner, D., Forster, C., Pinto Jr., O., and Calheiros, R.: Lightning activity in Brazilian thunderstorms during TROCCINOX: implications for NO_x production, Atmospheric Chemistry and Physics, 8, 921–953, http://www.atmos-chem-phys.net/8/921/2008/, 2008.

Huppert, R.: Theoretische und experimentelle Untersuchungen zum solaren I_0 Effekt, Diploma thesis, IUP Heidelberg, 2000.

Johnston, H. S.: Reduction of stratospheric ozone by nitrogen oxide catalysts from supersonic transport exhaust, Science, 173, 517–522, 1971.

Kraus, S.: DOASIS - DOAS Intelligent System, Ph.D. thesis, University Heidelberg, see http://www.iup.uniheidelberg.de/bugtracker/projects/doasis/, 2004a.

Kraus, S.: DOASIS - DOAS Intelligent System, see http://www.iup.uni-heidelberg.de/bugtracker/projects/doasis/, 2004b.

Kurucz, R., Furenhild, I., Brault, J., and Testermann, L.: Solar flux atlas from 296 to 1300 nm, National Solar Observatory Atlas No. 1, 1984.

McElroy, C. T.: Stratospheric nitrogen dioxide concentrations as determined from limb brightness measurements made on June 17th, J. Geophys. Res., 1983, 7075 – 7083, 1988.

Molina, L. T. and Rowland, F. S.: Stratospheric sink for chlorofluromethanes: chlorine atom catalyzed destruction of ozone, Nature, 249, 820–822, 1974.

Montzka, S., Butler, J., Hall, B., Mondell, D., and Elkins, J.: A decline in tropospheric organic bromine, Geophys. Res. Lett., 30, 1826–1829, 2003.

Noxon, J. F.: Nitrogen Dioxide in the Stratosphere and Troposphere measured by Ground-based Absorption Spectroscopy, Science, 189, 547–549, 1975.

Oelhaf, H., von Clarmann, T., Fergg, F., Fischer, H., Friedl-Vallon, F., Fritzsche, C., Piesch, C., Rabus, D., Seefeldner, M., and Volker, W.: Remote sensing of trace gases with a balloon-borne version of the Michelson interferometer for passive atmospheric sounding (MIPAS), ESA-SP-317, pp. 207–213, 10th ESA Symp. on European Rocket and Balloon Programmes and Related Research, 1991.

Pagsberg, P., E., B., Ratajczak, E., and Sillesen, A.: Kinetics of the gas phase reaction and the determination of the UV absorption cross sections of HONO, Chemical Physics Letters, p. 383 390, 1997.

Payan, S., Camy-Peyret, C., Lefevre, F., Jeseck, P., Hawat, T., and Durry, G.: First direct simultaneous HCl and $ClONO_2$ profile measurements in the Arctic vortex, Geophys. Res. Lett., 25, 2663–2666, 1998.

Payan, S., Camy-Peyret, C., Jeseck, P., Hawat, T., Pirre, M., Renard, J.-B., Robert, C., Lefevre, F., Kanzawa, H., and Sasano, Y.: Diurnal and nocturnal distribution of stratospheric NO_2 from solar and stellar occultation measurements in the Arctic vortex: Comparison with models and ILAS satellite measurements, J. Geophys. Res., 104, 21 585–21 593, 1999.

Pickering, K. E. e. a.: Vertical distributions of lightning NOx for use in regional and global chemical transport models, Atmos. Chem. Phys. Discuss., pp. 31 203–31 216, 1998.

Platt, U. and Stutz, J.: Differential Optical Absorption Spectroscopy (DOAS), Principle and Applications, ISBN 3-340-21193-4, Springer Verlag, Heidelberg, 2005.

Platt, U., Perner, D., and Pätz, H.: Simultaneous measurement of atmospheric CH_2O, O_3 and NO_2 by differential optical absorption, J. Geophys. Res., 84, 6329–6335, 1979.

Plumb, R. A.: A "tropical pipe" model of stratospheric transport, J. Geophys. Res., 101, 3957–3972, 1996.

Poeschl, U. e. a.: High Acetone Concentrations throughout the 0-12 km Altitude Range over the Tropical Rainforest in Surinam, J. Atmos. Chem., pp. 115–132, 2001.

Pommereau, J. and Goutail, F.: O3 and NO2 ground-based measurements by visible spectrometry during arctic winter and spring 1988, Geophys. Res. Lett., pp. 891–894, 1988.

Pope, F. D., Hansen, J. C., Bayes, K. D., Friedl, R. R., and Sander, S. P.: Ultraviolet Absorption Spectrum of Chlorine Peroxide, ClOOCl, J. Phys. Chem.A, 111, 4322–4332, 2007.

Ravishankara, A. R., Daniel, J. S., and Portmann, R. W.: Nitrous Oxide (N2O): The Dominant Ozone-Depleting Substance Emitted in the 21st Century, Science, p. 1176985, http://www.sciencemag.org/cgi/content/abstract/1176985v2, 2009.

Rodger, C. J. e. a.: Detection efficiency of the VLF World-Wide Lightning Location Network (WWLLN): initial case study, Ann. Geophys., pp. 3197–3214, 2006.

Rodgers, C.: Inverse methods for atmospheric sounding, World Scientific, Singapore, New Jersey, London, Hongkong, 2000.

Roscoe, H. K. and Pyle, J. A.: Measurements of solar occultation - The error in a naive retrieval if the constituent's concentration changes, J. Atmos. Chem., 5, 323–341, 1987.

Rothman, L. S., Jacquemart, D., Barbe, A., Chris Benner, D., Birk, M., Brown, L. R., Carleer, M. R., Chackerian, C., Chance, K., Dana, V., Devi, V. M., Flaud, J.-M., Gamache, R. R., Goldman, A., Hartmann, J.-M., Jucks, K. W., Maki, A. G., Mandin, J.-Y., Massie, S. T., Orphal, J., Perrin, A., Rinsland, C. P., Smith, M. A. H., Tennyson, J., Tolchenov, R. N., Toth, R. A., Vander Auwera, J., Varanasi, P., and Wagner, G.: The HITRAN 2004 Molecular Spectroscopic Database, J. Quant. Spec. and Rad. Transf., 96, 139–204, 2005.

Rozanov, A., Bovensmann, H., Bracher, A., Hrechanyy, S., Rozanov, V. V., Sinnhuber, M., Stroh, F., and Burrows, J. P.: NO_2 and BrO vertical profile retrieval from SCIAMACHY limb measurements: Sensitivity studies, Adv. Space Res., in press, 2005.

Salawitch, R., Weisenstein, D., Kovalenko, L., Sioris, C., Wennberg, P., Chance, K., Ko, M., and Mclinden, C.: Sensitivity of ozone to bromine in the lower stratosphere, Geophys. Res. Lett., 32, L05 811, 2005.

Salzmann, M. e. a.: Cloud system resolving model study of the roles of deep convection for photo chemistry in the TOGA COARE/CEPEX region, Atmos. Chem. Phys. Discuss., pp. 403–452, 2008.

Sander, S., Friedl, R., Ravishankara, A., Golden, D., Kolb, C. E., Kurylo, M., Huie, R., Orkin, V., Molina, M., Moortgat, G., and Finlayson-Pitts, B.: Chemical Kinetics and Photochemical Data for Use in Atmospheric Studies, JPL-Publication, 2003.

Sander, S., Friedl, R., Ravishankara, A., Golden, D., Kolb, C. E., Kurylo, M., Huie, R., Orkin, V., Molina, M., Moortgat, G., Keller-Rudek, H., and Finlayson-Pitts, P. B.: Chemical Kinetics and Photochemical Data for Use in Atmospheric Studies, JPL-Publication, 2006.

Schofield, R., Connor, B., Kreher, K., Johnston, P., and Rodgers, C.: The retrieval of profile and chemical information from ground-based UV-visible spectroscopic measurements, Journal of Quantative Spectroscopy and Radiative Transfer, 86, 115–131, 2004.

Sioris, C. E., Kurosu, T. P., Martin, R. V., and Chance, K.: Stratospheric and tropospheric NO_2 observed by SCIAMACHY: first results, Advances in Space Research, 34, 780–785, 2004.

Solomon, P., Barrett, J., Mooney, T., Connor, B., Parrish, A., and Siskind, D. E.: Rise and decline of active chlorine in the stratosphere, GEOPHYSICAL RESEARCH LETTERS, 33, 10.1029/2006GL027029, 2006.

Solomon, S., Garcia, R., and Ravishankara, A.: On the role of iodine in ozone depletion, J. Geophys. Res., 99, 20 491–20 499, 1994.

Stutz, J. and Platt, U.: Numerical Analysis and Estimation of the Statistical Error of Differential Optical Absorption Spectroscopy Measurements with Least-Squares methods, Appl. Opt., 35, 6041–6053, 1996.

Stutz, J. and Platt, U.: Improving long-path differential optical absorption spectroscopy with a quartz-fiber mode mixer, Applied Optics, 36, 1105–1115, 1997.

Stutz, J., Kim, E., Platt, U., Bruno, P., Perrino, C., and Febo, A.: UV-visible absorption cross-sections of nitrous acid, J. Geophys. Res., 105, 14,585–14,592, 2000.

Stutz, J., Alicke, B., and Neftel, A.: Nitrous acid formation in the urban atmosphere: Gradient measurements of NO2 and HONO over grass in Milan, Italy, J. Geophys. Res., 2002.

Trends, M. C. D.: ftp://ftp.cmdl.noaa.gov/ccg/co2/trends/co2_mm_mlo.txt, 2009.

Trishchenko, A. P., Luo, Y., Cribb, M. C., Li, Z., and Hamm, K.: Surface Spectral Albedo Intensive Operational Period at the ARM SGP Site in August 2002: Results, Analysis, and Future Plans, Thirteenth ARM Science Team Meeting Proceedings, Broomfield, Colorado, 2003.

Van de Hulst, H.: Light scattering by small particles, Dover publication, New York, 1981.

Voigt, S., Orphal, J., Bogumil, K., and Burrows, J. P.: The temperature dependence (203-293 K) of the absorption cross-sections of O_3 in the 230 - 850 nm region measured by Fourier-transform spectroscopy, J. of Photochemistry and Photobiology A: Chemistry, 143, 1–9, 2001.

von Clarmann, T. and Grabowski, U.: Elimination of hidden a priori information from remotely sensed profile data, Atmospheric Chemistry and Physics, 7, 397–408, http://www.atmos-chem-phys.net/7/397/2007/, 2007.

Wagner, T., Burrows, J. P., Deutschmann, T., Dix, B., von Friedeburg, C., Frieß, U., Hendrick, F., Heue, K.-P., Irie, H., Iwabuchi, H., Kanaya, Y., Keller, J., McLinden, C. A., Oetjen, H., Palazzi, E., Petritoli, A., Platt, U., Postylyakov, O., Pukite, J., Richter, A., van Roozendael, M., Rozanov, A., Rozanov, V., Sinreich, R., Sanghavi, S., and Wittrock, F.: Comparison of box-air-mass-factors and radiances for Multiple-Axis Differential Optical Absorption Spectroscopy (MAX-DOAS) geometries calculated from different UV/visible radiative transfer models, Atmospheric Chemistry and Physics, 7, 1809–1833, http://www.atmos-chem-phys.net/7/1809/2007/, 2007.

Wahner, A., Ravishankara, A. R., Sander, S. P., and Friedl, R. R.: Absorption cross section of BrO between 312 and 385 nm at 298 and 223 K, Chemical Physics Letters, 152, 507–512, 1988.

Weidner, F.: Development and Application of a Versatile Balloon-Borne DOAS Instrument for Skylight Radiance and Atmospheric Trace Gas Profile Measurements, Dissertation, Institut für Umweltphysik, Universität Heidelberg, 2005.

WMO: Scientific Assessment of Ozone depletion: 1998, *World Meteorological Organization Global Ozone Research and Monitoring Project, Report 44*, 1999.

WMO: Scientific Assessment of Ozone depletion: 2006, *World Meteorological Organization Global Ozone Research and Monitoring Project*, in preparation, 2006.

Wofsy, S., McElroy, M., and Yung, L.: The chemistry of atmospheric bromine, Geophys. Res. Lett., 2, 215–218, 1975.

Ya. B. Zeldovich, Y. P. R.: Physics of Shock Waves and High-Temperature Hydrodynamic Phenomena, Academic Press, New York and London, Academy of Sciences U.S.S.R. , Moscow, 1966.

Publications

Parts of this work will be published in (in preparation):

Kritten, L., Butz, A., M., Deutschmann, T., Dorf, M., Kühl, Prados-Roman, C., Pukite, J., Rozanov, A., Schofield, R. and Pfeilsticker, K.: Balloon-borne limb measurements of the diurnal variation of UV/vis absorbing radicals - a case study on NO_2 and O_3, Atmos. Meas. Tech., 2009.

Contributions have been made to the following articles:

Dorf, M., Butz, A., Camy-Peyret, C., Chipperfield, M. P., Kritten, L., and Pfeilsticker, K.: Bromine in the tropical troposphere and stratosphere as derived from balloon-borne BrO observations, Atmos. Chem. Phys., 8, 7265-7271,

Butz, A., Bösch, H., Camy-Peyret, C., Chipperfield, M. P., Dorf, M., Kreycy, S., Kritten, L., Prados-Román, C., Schwärzle, J., and Pfeilsticker, K.: Constraints on inorganic gaseous iodine in the tropical upper troposphere and stratosphere inferred from balloon-borne solar occultation observations, Atmos. Chem. Phys. Discuss., 9, 14645-14681

List of Figures

1.1 Vertical temperature profile of the Earth's atmosphere. Adopted from Brasseur and Solomon (1986). 14

1.2 Gaseous constituents of the atmosphere, with their order of magnitude in abundance and the number of different species occurring in that particular range. Adopted from: http://www.iup.uni-heidelberg.de/institut/studium/lehre/Atmosphaerenphysik/. 15

1.3 Stratospheric aerosol layer (Junge layer). Concentration of large particles (diameter $\geq 0.3\mu m$). Adopted from Chagnon and Junge (1961). 17

1.4 Global circulation. Adopted from http://rst.gsfc.nasa.gov/Sect14/Sect14_1c.html. 18

1.5 General circulation patterns in the atmosphere. For details see text. Adopted from WMO (1999). 19

1.6 ECMWF calculated time-height section of the monthly mean of zonal wind [m/s] at Teresina, northern Brazil, at 12 UT, in 2005 (courtesy of Katja Grunow, FU Berlin). 20

1.7 Summary of tropospheric/stratospheric characteristics and transitions thereof (symbolically shown as fade out of colored pattern).Θ, temperature lapse rate; Tmin, temperature minimum of profile; |T*|, amplitude of quasi-stationary zonal temperature anomaly; |T|, amplitude of tropical mean temperature seasonal cycle; QBO, quasi-biennial oscillation. Adopted from Fueglistaler et al. (2009). 21

1.8 Schematic of cloud processes and transport (left) and of zonal mean circulation (right) belonging to the TTL. Arrows indicate circulation, the black dashed line is the clear-sky level of zero net radiative heating (LZRH) and the black solid lines show isentropes in K, based on European Centre for Medium Range Weather Forecasts 40-year reanalysis (ERA-40). The relations between height, pressure and potential temperature are based on tropical annual mean temperature fields, with height values rounded to the nearest 0.5 km. The letters are explained in the text. Adopted from Fueglistaler et al. (2009). 22

1.9 In moist convection, air from near Earth's surface is rapidly transported upward and detrained into the UT. In this process, nitric acid (highly soluble) is efficiently scavenged while NO_x (insoluble) remains. NO_x is elevated by concurrent lightning NO production, resulting in high NO_x/HNO_3 ratios in the convective outflow region. After detrainment into the UT, NO_x is converted to HNO_3 by OH during the day and through reaction with NO_3, followed by hydrolysis of the N_2O_5 product, at night. The chemical evolution of the NO_x/HNO_3 ratio provides a unique indicator of the length of time that a sampled air mass has been in the UT after convection. Adopted from Bertram et al. (2007). 23

1.10 Possible reactions leading to HONO formation and loss in the upper troposphere 24

LIST OF FIGURES

1.11 Fractional contribution to odd oxygen (O_x) loss by catalytic cycles involving nitrogen (NO_x), hydrogen (HO_x), chlorine (ClO_x), bromine (BrO_x), and iodine (IO_x), calculated for March 1995, 32° N, using JPL-2002 kinetics (Sander et al., 2003). Br_y was derived from the breakdown of CH_3Br and halons plus additional 5 ppt representing $Br_y{}^{VSLS}$. The ClO_x curve represents loss from the ClO+O and $ClO+HO_2$ cycles, plus other minor cycles that involve ClO, but not BrO or IO. The BrO_x curve represents loss from the BrO+ClO and $BrO+HO_2$ cycles, plus other minor cycles that involve BrO, but not IO. Adopted from WMO (2006). 26

1.12 Diurnal variation of stratospheric nitrogen species. Temporal evolution of HNO_3 (upper left panel), NO_3 (middle left panel), NO (lower left panel), N_2O_5 (upper right panel), NO_2 (middle right panel) and SZA (lower right panel). The data are taken from a run of the LABMOS model of stratospheric chemistry on the 615 K potential temperature surface ($\approx$ 25 km). SZA = 90° is indicated by dotted vertical lines. 29

1.13 Altitude profiles of ClO over Hawaii (20° N) retrieved from spectra averaged over the indicated periods, illustrating the rapid rise in ClO from the early 1980s to the mid 1990s, the leveling off in the mid 1990s and a substantial decline by 2003 to about the 1992 level. Note that the 2003 and 1992 profiles almost overlap. Adopted from Solomon et al. (2006). 31

1.14 Overview about the chlorine (left) and bromine (right) primary source gases in 2004. Adopted from WMO (2006). 32

1.15 Measured trends for bromine (ppt) in the near-surface troposphere (lines) and stratosphere (squares). Global tropospheric bromine from methyl bromide as measured in ambient air and firn air (thin solid line - no correction has been made for tropospheric loss of CH_3Br) Butler et al. [1999] till 1998 and Montzka et al. [2003] past 1995; global tropospheric bromine from the sum of methyl bromide plus halons as measured in ambient air, archived air and firn air (thick solid line) Butler et al. [1999] and Fraser et al. [1999] till 1998 and Montzka et al. [2003] past 1995; and, bromine from CH_3Br and halons plus bromine from VSLS organic bromine compounds near the tropopause or transport of bromine bearing inorganic gases or bromine containing aerosols [Murphy and Thompson, 2000] across the tropopause (BryVSLS) [Salawitch et al., 2005; Pfeilsticker et al., 2000], assuming total contributions of 3, 5, or 7 ppt of these species to Br_y (thin dotted lines). Total inorganic bromine derived from stratospheric measurements of BrO and photochemical modeling that accounts for BrO/Bry partitioning from slopes of Langley BrO observations above balloon float altitude (filled squares) and lowermost stratospheric BrO measurements (open squares). Bold/faint error bars correspond to the precision/accuracy of the estimates, respectively. The years indicated on the abscissa are sampling times for tropospheric data. For stratospheric data, the date corresponds to the time when the air was last in the troposphere, i.e. sampling date minus estimated mean time in stratosphere. Preindustrial levels of CH_3Br were (5.8 $\pm$ 0.3) ppt [Saltzman et al., 2004] in the southern hemisphere and 0 ppt for the halons [Reeves et al., 2005]. Adopted from an update of Dorf et al. (2006). 33

2.1 Sunset at the measurement side, in Teresina, Brasil: Rayleigh scattering causes the blue hue of the daytime sky and the reddening of the sun at sunset, the scattering occurring on cloud droplets is explained by Mie theory. 37

2.2 Actinic fluxes simulated with McArtim (Deutschmann, 2008) for 5 km (green), 29 km (black) and 69 km (red) height on a clear day. Left panel: On a logarithmic scale, highlighting the absorption between 200 and 300 nm, which is mainly due to ozone and oxygen. Right panel: On a linear scale, highlighting the visible wavelength range, where the decrease at 5 km is mainly due to Mie scattering. 40

LIST OF FIGURES

2.3 Simulated radiative transfer events of balloon-borne limb scattered skylight measurements with our mini-DOAS instrument in 34 km altitude and elevation of the telescope $\alpha = -2°$. The coloured dots denote photon - matter interaction, Rayleigh scattering (red), Mie scattering (green), absorption (blue) and ground scattering (yellow). 41

3.1 Sketch of the mini-DOAS instrument components. A detailed description is given in the text. 43

3.2 Sketch of the optics governed by the telescope and the fibers. Note, that the light throughput D is independent of the lens for isotropic radiation, while the FOV is altered by the use of a lens. [Sketch from Bodo Werner, personal communication]. 44

3.3 Left panel: Total noise as a function of co-added spectra at the 80% illumination level of the CCD array detector for skylight. Right panel: Peak to peak residual of the same spectra. 47

3.4 Left panel: Signal as a function of total integration time at constant illumination. Right panel: The ratio of signal to total noise as a function of total integration time. 47

3.5 Illustration of the mini-DOAS attitude with respect to the earth coordinate system, by introducing an additional coordinate system in which the gondola is fixed (gondola coordinate system), where moving of the telescope is only governed by the mini-DOAS instruments controll system. The resulting angles α^{**} (between $x_{gondola}$ and x_{earth}), α^{*} (between $x_{gondola}$ and the telescope optical axis) and α (between x_{earth} and the telescope optical axis), which is the sum of both and an important input for RTM. 48

3.6 Viewing geometry of balloon borne Limb scattered skylight measurements during balloon ascent (left panel) and during float (right panel). 49

3.7 Sketch of the triangulation with the laser pointer to fix the elevation angle of the telescope prior to the flight, d, x_1, x_2, α^* as explained in the text. 50

4.1 Schematic drawing of the procedure applied to retrieve time series of trace gas profiles and chemical information from mini-DOAS measurements. Input and output parameter are shown in yellow, modeling is colored blue and retrievals are shown in green (for some boxes this categorization is ambiguous). 53

4.2 Sample DOAS evaluation of O_3 (left panel), which is performed in the wavelength interval $490 - 520$ nm and of NO_2 (right panel), which is performed the in the wavelength interval $435 - 460$ nm. Shown is the optical density of the absorbance of the trace gases and the Ring effect. The lowest two traces show the measured (red line) and the reference (black line) spectra. The panels above illustrate the remaining residuals of the fitting procedure. The red lines indicate the spectral absorption and the black lines the sum of the spectral absorption and the residual. 61

4.3 Sample DOAS evaluation of O_4, which is performed in the wavelength interval $465 - 490$ nm. Shown is the optical density of the absorbance of the trace gases and Ring. The lowest two traces show the measured (red line) and the reference (black line) spectra. The panels above illustrate the remaining residuals of the fitting procedure. The red lines indicate the spectral absorption and the black lines the sum of the spectral absorption and the residual. 62

4.4 Sample DOAS evaluation of HONO (left panel), which is performed in the wavelength interval $348 - 392$ nm and of BrO (right panel), which is performed in the wavelength interval $346 - 359$ nm. Shown is the optical density of the absorbance of the trace gases and the Ring. The lowest two traces show the measured (red line) and the reference (black line) spectra. The panels above illustrate the remaining residuals of the fitting procedure. The red lines indicate the spectral absorption and the black lines the sum of the spectral absorption and the residual. 63

4.5　BoxAMFs ($\mathbf{K}_{i,j}$) for the NO_2 concentration profile, for Limb scans, recorded at 35 km altitude in series from 0.5° to -5.5° elevation angle from aboard the LPMA/IASI payload on June 30, 2005. .　65

4.6　Left panel: Modeled (red) and measured (black) Limb radiances, with the modeled values shifted to measured values (blue). Right panel: Wavelength dependency of the retrieved $\Delta\alpha$. . . . .　66

4.7　Effect of a Gaussiantype FOV on the ΔBoxAMFs. Left panel: Rectangular instrumental FOV and corresponding BoxAMFs. Right panel: Gaussiantype FOV for a variance $\sigma = 0.3°$ and corresponding BoxAMFs. . . . .　67

4.8　Example time weighting matrix C for the Limb observation from aboard the MIPAS-B payload over Teresina on June 14, 2005. . . . .　69

4.9　Logarithm of the kernel $\mathbf{K}_{i,j}$ for the retrieved NO_2 concentration profile at $T_k = 11:30$ [UT]. The example is for Limb scanning measurements at 35 km altitude with subsequent measurements in steps of 0.5° from $\alpha = 0.5°$ to $\alpha = -5.5°$ elevation angle from aboard the LPMA/IASI payload on June 30, 2005. . . . .　70

4.10　Measured (black) and forward modeled (red) $\Delta SCDs$ for the observation from aboard LPMA/IASI payload on June 30, 2005. The difference between modeled and measured $\Delta SCDs$ is shown in green. . . . .　71

4.11　L-curve estimation of the a priori error $\mathbf{S_a}$. . . . .　73

4.12　Degrees of freedom of the retrieval (trace of the averaging kernels matrix A) plotted as a function of the HWHM (h). This curve has been calculated for the retrieval of NO_2 from Limb scanning on June 30, 2005. . . . .　74

4.13　Left panel: Concentration of NO_2 retrieved on a coarse retrieval grid. Right panel: Averaging kernels from the retrieval on a coarse grid. . . . .　75

4.14　Gradient of the NO_2 concentration over time, as modeled with Labmos. . . . .　76

4.15　Two dimensional averaging kernel of the of the NO_2 concentration retrieval at 33 km, at 14 UT on June 30, 2005. . . . .　78

4.16　Upper left panel: Extreme cases (type I around 13:15 UT and type II around 14:15 UT) of α^{**} oscillation of the MIPAS-B gondola, as recorded by attitude control system. Lower left panel: Concentration of NO_2 retrieved from $\Delta SCDs$ expected on a calm flight (green) and on a gondola undergoing type I oscillation (black). Lower middle panel: Concentration of NO_2 retrieved from $\Delta SCDs$ expected on a calm flight (green) and on a gondola undergoing type II oscillation (black). Right panel: Differences of the retrieved profiles during both oscillation types shown on the left panel (red), envelope of the differences defined as an upper limit for $\mathbf{S_{osci}}$ (green error bars) and noise error $\mathbf{S_{noise}}$ (black error bars) for comparison. . . . .　80

5.1　Left panel: Satellite picture of north-east South America, with a mark at the measurement site. Adopted from: http://maps.google.de/maps. Right panel: Photography of the LPMA/DOAS payload during launch. Visible is the gondola hanging below the two auxiliary balloons, the much bigger main balloon is higher above. . . . .　83

5.2　Relative humidity (blue) and temperature (black) as a function of altitude, measured by sonding at Teresina on June 17, 2005. The approximate height of the cold point tropopause is shown in purple dots (≈ 17 km). . . . .　84

LIST OF FIGURES

5.3 Left panel: MIPAS-B gondola with the position of the mini-DOAS instrument indicated by a green box [Sketch from Hans Nordmeyer, personal communication]. Right panel: Flight profile of the MIPAS-B gondola on June 13, 2005. Altitude (black) and SZA (red) versus time. Part I and part II denote periods of different float altitudes. 85

5.4 Logarithm of the weighting function $\mathbf{K}_{i,j}$ matrix elements for the retrieved NO_2 concentration profile at t_k = 14 [UT]. The example is for limb scanning measurements at around 33 km altitude with subsequent measurements in steps of 0.5° from α=-0.5° to α=-6° elevation angle from aboard the MIPAS-B payload, 2005. 86

5.5 Measured (black), and simulated (red, blue) O_3 ΔSCDs from limb scanning measurements from aboard the MIPAS-B gondola on June 14, 2005. The forward modeled retrieved profiles are shown in red and the difference to the measured ΔSCDs (residual of the fit) is shown in green. Forward modeled $a\ priori$ profiles (blue) are plotted for comparison. 87

5.6 Measured (black), and simulated (red, blue) NO_2 ΔSCDs from limb scanning measurements from aboard the MIPAS-B gondola on June 14, 2005. The forward modeled retrieved profiles are shown in red and the difference to the measured ΔSCDs (residual of the fit) is shown in green. Forward modeled $a\ priori$ profiles (blue) are plotted for comparison. 88

5.7 Measured (black), and simulated (red, blue) BrO ΔSCDs from limb scanning measurements from aboard the MIPAS-B gondola on June 14, 2005. The forward modeled retrieved profiles are shown in red and the difference to the measured ΔSCDs (residual of the fit) is shown in green. Forward modeled $a\ priori$ profiles (blue) are plotted for comparison. 89

5.8 Measured (black), and simulated (red, blue) O_4 ΔSCDs from limb scanning measurements from aboard the MIPAS-B gondola on June 14, 2005. The forward modeled retrieved profiles are shown in red and the difference to the measured ΔSCDs (residual of the fit) is shown in green. Forward modeled $a\ priori$ profiles (blue) are plotted for comparison. 90

5.9 Measured (black), and simulated (red, blue) HONO ΔSCDs from limb scanning measurements from aboard the MIPAS-B gondola on June 14, 2005. The forward modeled retrieved profiles are shown in red and the difference to the measured ΔSCDs (residual of the fit) is shown in green. Forward modeled $a\ priori$ profiles (blue) are plotted for comparison. 91

5.10 Retrieval of O_3 from balloon-borne measurements on board MIPAS-B payload on June 14, 2005. Upper panel: O_3 concentration versus altitude and time. Lower panel: Area of the averaging kernels versus altitude and time. 92

5.11 Averaging kernels of the second limb profile (indicated in Figure 5.10 by a black line)versus altitude. 93

5.12 Retrieval of NO_2 from balloon-borne measurements on board MIPAS-B payload on June 14, 2005. Upper panel: NO_2 concentration versus altitude and time. Lower panel: Area of the averaging kernels versus altitude and time. 94

5.13 Averaging kernels of the second limb profile (indicated in Figure 5.12 by a black line) versus altitude. 95

5.14 Retrieval of BrO from balloon-borne measurements on board MIPAS-B payload on June 14, 2005. Upper panel: BrO concentration versus altitude and time. Lower panel: Area of the averaging kernels versus altitude and time. 96

5.15 Averaging kernels of the second limb profile (indicated in Figure 5.14 by a black line) versus altitude. 97

5.16 Retrieval of HONO from balloon-borne measurements on board MIPAS-B payload on June 14, 2005. Upper panel: HONO concentration versus altitude and time. Lower panel: Area of the averaging kernels versus altitude and time. 98

5.17 Averaging kernels of the second limb profile (indicated in Figure 5.16 by a black line) versus altitude. 99

5.18 Left panel: Concentration profile of O_3, retrieved from balloon-borne measurements on board MIPAS-B payload on June 14, 2005 (black) and from in-situ sonde in the vicinity of the balloon flight (red), smoothed with the averaging kernels of the mini-DOAS retrieval(green). Right panel: Averaging kernels of the mini-DOAS retrieval of O_3. 100

5.19 High clouds as seen from the Mipas-B star camera looking to NNE at about 40 min before sunrise (9:15 UT) on June 14 2005. Overshooting cloud tops are at about 12 km, a thin cirrus cloud is at about 14 km. Higher up in the clear stratosphere some stars are still visible. 101

5.20 Left panel: Azimuthal viewing direction (black lines) of the mini-DOAS instrument during MIPAS-B flight (times given in UT). Right panel: Map of north-eastern Brazil with real-time locations of lightning discharges from the World-Wide Lightning Location Network (WWLLN) for the same day. The colour codes time in UT. The green star shows the location of the measurements. 102

5.21 Left panel: Observed NO_2 (black), calculated HONO, which is used as *a priori* for the inversion (red) and retrieved HONO (green), height of the tropopause (blue). Right panel: Concentration of OH (black, calculated by equation 5.3) which is necessary to explain the measured amount of HONO and expected concentration of OH (red) (Salzmann, 2008). 103

5.22 Flight profile of the LPMA/DOAS gondola on June 17, 2005. Altitude (black) and SZA (red) versus Universal Time. 104

5.23 Logarithm of the weighting function $K_{i,j}$ matrix elements for the retrieved NO_2 concentration profile at $t_k = 19:42$ [UT]. The example shows measurements during balloon ascent from 15 to 31 km altitude with a telescope elevation of $\alpha = -2°$ from aboard the LPMA/DOAS payload. . . . 105

5.24 Measured (black), and simulated (red, blue) O_3 ΔSCDs from Limb scanning measurements from aboard the LPMA/DOAS gondola on June 17, 2005. The forward modeled retrieved profiles are shown in red and the difference to the measured ones (residual of the fit) is shown in green. Forward modeled *a priori* profiles (blue) are plotted for comparison. 106

5.25 Measured (black), and simulated (red, blue) NO_2 ΔSCDs from Limb scanning measurements from aboard the LPMA/DOAS gondola on June 17, 2005. The forward modeled retrieved profiles are shown in red and the difference to the measured ΔSCDs (residual of the fit) is shown in green. Forward modeled *a priori* profiles (blue) are plotted for comparison. 107

5.26 Measured (black), and simulated (red, blue) BrO ΔSCDs from Limb scanning measurements from aboard the LPMA/DOAS gondola on June 17, 2005. The forward modeled retrieved profiles are shown in red and the difference to the measured ΔSCDs (residual of the fit) is shown in green. Forward modeled *a priori* profiles (blue) are plotted for comparison. 108

5.27 Measured (black), and simulated (blue, pink, cyan) O_4 ΔSCDs from Limb scanning measurements from aboard the LPMA/DOAS gondola on June 17, 2005. The different simulated ΔSCDs are obtained by applying different aerosol extinction profiles (figure 5.28) for the RTM. 109

5.28 Aerosol extinction profiles as applied for the RTM of Limb scanning measurements from aboard the LPMA/DOAS gondola, for 490nm. 110

5.29 Left panel: Logarithm of the ratio of measured intensity to reference intensity compared to the ratio of modeled intensity to reference intensity. Right panel: Retrieved aerosol extinction profile. [Plots from Tim Deutschmann, personal communication] 111

5.30 Retrieval of O_3 from balloon-borne measurements on board LPMA/DOAS payload on June 17, 2005. Upper panel: O_3 concentration versus altitude and time. Lower panel: Area of the averaging kernels versus altitude and time. 112

LIST OF FIGURES

5.31 Averaging kernels of the second limb profile (indicated in Figure 5.30 by a black line) versus altitude. 113

5.32 Retrieval of NO_2 from balloon-borne measurements on board LPMA/DOAS payload on June 17, 2005. Upper panel: NO_2 concentration versus altitude and time. Lower panel: Area of the averaging kernels versus altitude and time. 114

5.33 Averaging kernels of the second limb profile (indicated in Figure 5.32 by a black line) versus altitude. 115

5.34 Retrieval of BrO from balloon-borne measurements on board LPMA/DOAS payload on June 17, 2005. Upper panel: BrO concentration versus altitude and time. Lower panel: Area of the averaging kernels versus altitude and time. 116

5.35 Averaging kernels of the second limb profile (indicated in Figure 5.34 by a black line) versus altitude. 117

5.36 Left upper and lower panel: Concentration profile of O_3, from balloon-borne measurements of scattered light (black), direct sunlight (red) and from in-situ measurements by an electrochemical cell aboard the gondola (green). Right upper and lower panel: Averaging kernels of the scattered light retrieval. 118

5.37 Left upper and lower panel: Concentration profile of NO_2, retrieved from balloon-borne measurements of scattered light (black) and direct sun (red). Right upper and lower panel: Averaging kernels of the scattered light retrieval. 119

5.38 Left upper and lower panel: Concentration profile of BrO, retrieved from balloon-borne measurements of scattered light (black) and direct sun (red). Right upper and lower panel: Averaging kernels of the scattered light retrieval. 120

5.39 Flight profile of the IASI gondola on June 30, 2005. Altitude (black) and SZA (red) versus Universal Time. 121

5.40 Measured (black) and forward modeled (red) O_3 ΔSCDs from Limb scanning measurements from aboard the LPMA/IASI gondola on June 30, 2005. The difference between forward modeled and measured ΔSCDs is shown in green. 122

5.41 Measured (black) and forward modeled (red) NO_2 ΔSCDs from Limb scanning measurements from aboard the LPMA/IASI gondola on June 30, 2005. The difference between forward modeled and measured ΔSCDs is shown in green. 123

5.42 Measured (black) and forward modeled (red) BrO ΔSCDs from Limb scanning measurements from aboard the LPMA/IASI gondola on June 30, 2005. The difference between forward modeled and measured ΔSCDs is shown in green. 124

5.43 Measured (black) and forward modeled (red) O_4 ΔSCDs from Limb scanning measurements from aboard the LPMA/IASI gondola on June 30, 2005. The difference between forward modeled and measured ΔSCDs is shown in green. 125

5.44 Retrieval of O_3 from balloon-borne measurements on board LPMA/IASI payload on June 14, 2005. Upper panel: O_3 concentration versus altitude and time. Lower panel: Area of the averaging kernels versus altitude and time. 126

5.45 Averaging kernels of the second limb profile (indicated in Figure 5.44 by a black line) versus altitude. 127

5.46 Retrieval of NO_2 from balloon-borne measurements on board LPMA/IASI payload on June 14, 2005. Upper panel: NO_2 concentration versus altitude and time. Lower panel: Area of the averaging kernels versus altitude and time. 128

5.47 Averaging kernels of the second limb profile (indicated in Figure 5.46 by a black line) versus altitude. 129

5.48 Retrieval of BrO from balloon-borne measurements on board LPMA/IASI payload on June 14, 2005. Upper panel: NO_2 concentration versus altitude and time. Lower panel: Area of the averaging kernels versus altitude and time. 130

5.49 Averaging kernels of the second limb profile (indicated in Figure 5.48 by a black line) versus altitude. 131

5.50 Left panel: Retrieved NO_2 concentration, from SCIAMACHY measurements in orbit 17427 by the IUP Bremen (red) and from balloon-borne measurements on board LPMA/IASI payload by the IUP Heidelberg (black), at SZA = 34° on June 30, 2005. Right panel: Left side: Averaging kernels of the retrieval of NO_2 from SCIAMACHY measurements by the IUP Bremen (red) and from balloon-borne measurements from the IUP Heidelberg (black). Middle: Backus-Gilbert spread of the averaging kernels. Right side: Degrees of freedom of the retrieval. 133

5.51 Retrieved concentration, from SCIAMACHY measurements in orbit 17427 by the IUP Bremen (red) and from balloon-borne measurements on board LPMA/IASI payload by the IUP Heidelberg (black), at SZA = 34° on June 30, 2005. Left panel: O_3. Right panel: BrO. 133

5.52 Reaction scheme of NO_y. The reactions in the orange boxes are considered in our simplified model. 134

5.53 NO_x/NO_y species measured by MIPAS-B on June 14, 2005. Upper panels from left to right: $ClONO_2$, HNO_3 and N_2O_5. Lower panels: N_2O and NO_2. 135

5.54 Photolysis rate of N_2O_5 ($J_{N_2O_5}$) for different SZAs (coloured lines), calculated by McArtim actinic fluxes using absorption cross sections from JPL 2006 (Sander et al., 2006). 135

5.55 Inverse lifetime of N_2O_5 due to photolysis for an SZA range from 72 to 29°, using the molecular and kinetic data from JPL 2006 (Sander et al., 2006). 137

5.56 Time series of measured (black) and modeled (gray) NO_2 concentration for different altitudes. The uncertainty range of the modeled data is due to the combined uncertainty for cross section and quantum yield. 138

8.1 Subsequent concentration profiles of O_3, retrieved from mini-DOAS measurements on the Mipas-B payload on June 14, 2005 144

8.2 Subsequent concentration profiles of NO_2, retrieved from mini-DOAS measurements on the Mipas-B payload on June 14, 2005 145

8.3 Subsequent concentration profiles of BrO, retrieved from mini-DOAS measurements on the Mipas-B payload on June 14, 2005 146

8.4 Subsequent concentration profiles of HONO, retrieved from mini-DOAS measurements on the Mipas-B payload on June 14, 2005 147

8.5 Subsequent concentration profiles of O_3, retrieved from mini-DOAS measurements on the LPMA/DOAS payload on June 17, 2005 148

8.6 Subsequent concentration profiles of NO_2, retrieved from mini-DOAS measurements on the LPMA/DOAS payload on June 17, 2005 149

8.7 Subsequent concentration profiles of BrO, retrieved from mini-DOAS measurements on the LPMA/DOAS payload on June 17, 2005 150

8.8 Concentration profile of O_3, retrieved from balloon-borne measurements on board LPMA/IASI payload on June 14, 2005 (black) and *a priori* profile (red). 151

8.9 Concentration profile of NO_2, retrieved from balloon-borne measurements on board LPMA/IASI payload on June 14, 2005 (black) and *a priori* profile (red). 152

LIST OF FIGURES

8.10 Concentration profile of BrO, retrieved from balloon-borne measurements on board LPMA/IASI payload on June 14, 2005 (black) and *a priori* profile (red). 153

List of Tables

3.1 Compendium of balloon-borne mini-DOAS measurements. The last eight flights were conducted within the present thesis. 51

4.1 Compendium of molecular absorption cross sections used for the spectral analysis. 60

4.2 The second derivative operator **L**. 75

VDM Verlagsservicegesellschaft mbH

Die VDM Verlagsservicegesellschaft sucht für wissenschaftliche Verlage abgeschlossene und herausragende

Dissertationen, Habilitationen, Diplomarbeiten, Master Theses, Magisterarbeiten usw.

für die kostenlose Publikation als Fachbuch.

Sie verfügen über eine Arbeit, die hohen inhaltlichen und formalen Ansprüchen genügt, und haben Interesse an einer honorarvergüteten Publikation?

Dann senden Sie bitte erste Informationen über sich und Ihre Arbeit per Email an *info@vdm-vsg.de*.

Sie erhalten kurzfristig unser Feedback!

VDM Verlagsservicegesellschaft mbH
Dudweiler Landstr. 99   Telefon +49 681 3720 174
D - 66123 Saarbrücken   Fax   +49 681 3720 1749

www.vdm-vsg.de

Die VDM Verlagsservicegesellschaft mbH vertritt

Printed by Books on Demand GmbH, Norderstedt / Germany